SpringerBriefs in Earth System Sciences

Series Editors

Gerrit Lohmann, Universität Bremen, Bremen, Germany

Justus Notholt, Institute of Environmental Physics, University of Bremen, Bremen, Germany

Jorge Rabassa, Labaratorio de Geomorfología y Cuaternar, CADIC-CONICET, Ushuaia, Tierra del Fuego, Argentina

Vikram Unnithan, Department of Earth and Space Sciences, Jacobs University Bremen, Bremen, Germany

SpringerBriefs in Earth System Sciences present concise summaries of cutting-edge research and practical applications. The series focuses on interdisciplinary research linking the lithosphere, atmosphere, biosphere, cryosphere, and hydrosphere building the system earth. It publishes peer-reviewed monographs under the editorial supervision of an international advisory board with the aim to publish 8 to 12 weeks after acceptance. Featuring compact volumes of 50 to 125 pages (approx. 20,000—70,000 words), the series covers a range of content from professional to academic such as:

- A timely reports of state-of-the art analytical techniques
- bridges between new research results
- snapshots of hot and/or emerging topics
- literature reviews
- in-depth case studies

Briefs are published as part of Springer's eBook collection, with millions of users worldwide. In addition, Briefs are available for individual print and electronic purchase. Briefs are characterized by fast, global electronic dissemination, standard publishing contracts, easy-to-use manuscript preparation and formatting guidelines, and expedited production schedules.

Both solicited and unsolicited manuscripts are considered for publication in this series.

Jack J. Middelburg

Thermodynamics and Equilibria in Earth System Sciences: An Introduction

 Springer

Jack J. Middelburg
Department of Earth Sciences
Utrecht University
Utrecht, The Netherlands

ISSN 2191-589X   ISSN 2191-5903 (electronic)
SpringerBriefs in Earth System Sciences
ISBN 978-3-031-53406-5   ISBN 978-3-031-53407-2 (eBook)
https://doi.org/10.1007/978-3-031-53407-2

This Springer imprint is published by the registered company Springer Nature Switzerland AG
The registered company address is: Gewerbestrasse 11, 6330 Cham, Switzerland

Paper in this product is recyclable.

Preface

Thermodynamics deals with energy conversions and predicts the directions of these energy transformations. Thermodynamics is needed to understand many processes on Earth, be they physical, chemical, or biological. Thermodynamics is critical to study the atmosphere (lapse rate, foehn winds, circulation), hydrosphere (latent and sensible heat, pressure dependence of freezing/boiling points), geosphere (geothermal gradients, mineral stability), and the biosphere (redox zonation, evolution of biogeochemical cycles).

Thermodynamics is based on two major laws. The first law concerns the conservation of energy and the second law states that energy is increasingly being dispersed. These two principal laws of thermodynamics are general and simple, both in words and equations, but their applications may be challenging. Citing the Feynman lectures on Physics (Feynman et al., 1963): *"Thermodynamics is a rather difficult and complex subject when we come to apply it, and it is not appropriate for us to go very far into the applications in this course. ...The subject of thermodynamics is complicated because there are so many different ways of describing the same thing"* It is also challenging to teach to undergraduate students, because some might better grasp it with a rigorous approach involving calculus and derivation of each equation, while others would understand the topic better using everyday examples and limiting the use of mathematics.

This introduction to thermodynamics and equilibria is part of the first-year chemistry class for Earth (System) Science students at Utrecht University. It aims to prepare students for more advanced second-year classes in physical chemistry, mineralogy, and petrology and to provide the basic concepts of relevance for atmospheric, marine, climate, and environmental sciences. Mathematics is limited to the inevitable and notations are kept as simple as possible to maintain accessibility of the text. This introduction is limited to ideal, equilibrium systems, neglects the temperature and pressure dependence of enthalpy, entropy, and heat capacities, and does not explicitly mention the reversible or irreversible nature, or the path, of system changes. These and other simplifications at the expense of scientific rigor and comprehensiveness are hopefully compensated by students' improved understanding how to use thermodynamics for problems in Earth System Science.

This concise, accessible introduction was written during a one-month workation in the Ardennes (May 2023) with the aim to have a revised first-year course ready for the 2024 spring semester. The author acknowledges Jeannot Trampert and in particular Lubos Polerecky for critical feedback on the first draft, teaching assistants Chris Bil and Thomas Sanders for checking the second draft, and Ton Markus for upgrading the quality of the figures.

Utrecht, The Netherlands   Jack J. Middelburg

Contents

**Part II   Equilibria: Solutions, Minerals, Acid-Base and Redox
        Reactions**

Part I
Thermodynamics

Chapter 1
Introduction

Abstract This chapter introduces the macroscopic approach to matter and defines thermodynamic systems and their exchange with the surroundings. The state variables temperature and pressure are presented and linked via the ideal gas law (an equation of state for gas).

Keywords Partial pressure · Temperature · Ideal gas law · Thermodynamic system

Thermodynamics is a scientific discipline firmly rooted in physics, chemistry and engineering with implications for the universe and thus biology, earth sciences and astronomy. It describes the behavior of matter on a macroscale, sets the rules for conversions of energy from one form into another, and governs the direction of processes. Engineers and physicists applying thermodynamics focus on energy conversions and conservation to develop efficient engines, refrigerators, or heat pumps. Chemists use thermodynamics to study chemical reactions and phase equilibria. Thermodynamics is key to understand many processes on Earth including adiabatic compression of air (Fohn winds), the lapse rate due to expansion of air, geothermal gradients in the inner Earth, the effect of pressure on freezing and boiling points, the strength of atmospheric circulation, the stability of minerals as a function of pressure, temperature, and composition, redox zonation in marine sediments and the evolution of biogeochemical cycles.

Thermodynamics is a science about the macroscopic world, i.e., the human scale and larger. Most of the thermodynamic concepts have been developed, and can be understood, without knowledge of the molecular structure. However, sometimes it is easier understood if we use our knowledge on the properties of individual molecules.

Thermodynamics is based upon a few statements, i.e., thermodynamic laws, that are based on years of observations and multiple clever experiments connected through theory. Most chemical thermodynamics textbooks present four laws, while engineering treatments are often limited to the first and second laws. *The first law is the conservation of energy*: energy can be converted from one form into another, but overall, no energy is lost or gained. *The second law is concerned with spontaneous change* and states that energy is, overall and over time, increasingly being dispersed.

© The Author(s) 2024

J. J. Middelburg, *Thermodynamics and Equilibria in Earth System Sciences: An Introduction*, SpringerBriefs in Earth System Sciences,
https://doi.org/10.1007/978-3-031-53407-2_1

Alternatively formulated, the second law reads that spontaneous processes are those which increase the entropy of the universe. The term entropy will be introduced and explained later.

1.1 Macroscopic Approach, Yet Everything is Made of Atoms

Chemists deal with atoms, ions, and molecules and aim to explain the properties of matter and predict the synthesis of compounds using atom-level knowledge. Although thermodynamics is a core subject in academic chemistry curricula, it deals with macroscopic systems, i.e., large amounts of matter rather than a few molecules. Matter exists in three phases: solid, liquid or gas that differ in the number of particles (atoms, molecules, ions) per volume, hence the distance and interactions among particles, and their spatial orientations and distribution (Fig. 1.1). In a gas molecules are far apart and move randomly through a volume that is primarily empty space. A gas is thus homogenous, fluid, and compressible. Moreover, for an ideal gas the identity and interaction of the particles can be ignored at the macroscopic level. In a liquid the particles are close to each other but are free to move relative to each other, thus a liquid is fluid but less compressible. In a solid constituent, atoms, ions, and molecules are generally close to each other and in fixed positions. Moreover, their identity governs the physical and chemical properties of the solid. Pure substances, matter with a homogenous and definite chemical composition, may exist as a solid, liquid or gas depending on the external conditions (temperature, pressure). For instance, at the Earth surface, H_2O, water, may exist as solid (ice), liquid (water) or gas (water vapor). A single phase may contain multiple substances, e.g., air consists of nitrogen, oxygen, carbon dioxide, etc.

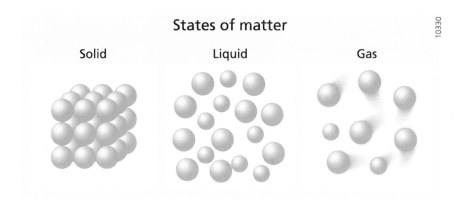

Fig. 1.1 The most common states of matter: solid, liquid and gas

Table 1.1 Pressure units and their relation to SI unit Pa

1 Pa	1 Nm^{-2}
1 bar	10^5 Pa
1 mm Hg	133.32 Pa
1 torricille or torr	133.32 Pa
1 atm	101.325 Pa

Thermodynamic theory is general and applies to solid, liquid, and gas phases. However, most thermodynamics has been developed based on the behavior of gas. Consider a container filled with a gas. Gas exerts a pressure on the container walls, and this can be understood in terms of randomly moving particles that collide with the wall. The result of these collisions is a force perpendicular to the wall. Gas pressure (P) is thus defined as:

$$P = \frac{F}{A}, \tag{1.1}$$

where F = force (N) and A = area (m^2). Pressure acts equally in all directions in a gas and is thus a scalar quantity. Pressure has the unit Nm^{-2} or Pa (the SI unit), but multiple other units are commonly used (Table 1.1).

Because of the large amount of space between the molecules in a gas (Fig. 1.1), and thus limited interactions, gases with different compositions will mix well. If they do not react, the total pressure of the gas mixture is the sum of the partial pressure of each gas contributing to the mixture. The partial pressure of gas A (P$_A$) is the pressure it would have if it were alone. *Dalton's law* states that the total pressure (P$_{tot}$) for a mixture of gases A, B, C, is given by:

$$P_{tot} = P_A + P_B + P_C \tag{1.2}$$

For instance, for air (78% N$_2$, 21% O$_2$) with a total pressure of 1 atm, the partial pressure of nitrogen and oxygen gases would be 0.78 and 0.21 atm, respectively.

The particles moving freely in a gas have kinetic energy. Temperature is a measure of how much kinetic energy each particle has on average. The higher the temperature, the more energy a system has, all other factors being the same. Temperature is not a form of energy, but a measurable parameter to compare amounts of energy of different systems. Most solids, liquids, and gases expand roughly linearly with increasing temperature because temperature governs the kinetic energy of particles. The expansion of liquids such as mercury and alcohol are commonly used to quantify temperature using the freezing (0) and boiling (100) points of water for calibration. This centigrade scale has been superseded by the closely related Celsius scale anchored to the triple point of water (0.01 °C, see Sect. 4.3) at which ice, water and gaseous water are at equilibrium. The Celsius scale is widely used but the scale is arbitrary: it is based on water and its macroscopic phases. The thermodynamic, or absolute, temperature scale is independent of the substance used in the thermometer and goes down to the minimum possible temperature of -273.15 °C. The absolute scale, or Kelvin scale, is expressed in K (without a degree sign). Temperature in degrees Celsius and Kelvins are related:

$$K = °C + 273.15 \qquad (1.3)$$

This Kelvin scale should be used in thermodynamics.

Having defined pressure (P) and temperature (T). We can define two reference states: standard temperature and pressure (STP) refers to $P = 1$ bar and $T = 273.15$ K $= 0.0$ °C, while standard ambient temperature and pressure (SATP) refers to $T = 298.15$ K $= 25$ °C and 1 bar for P (1 bar $= 10^5$ Pa).

1.2   Ideal Gas Law

Experimental studies on the physical properties of gases have resulted in a few empirical laws that were eventually combined into what we call today the ideal gas law. Boyle (1627–1691) studied the relationship between the pressure and volume of a fixed amount of gas at constant temperature and observed that the product of pressure and volume is constant. *Boyle's law* states that the volume (V) of a given amount (e.g., mass) of a gas is inversely proportional to pressure (P) when the temperature is constant:

$$V \propto \frac{1}{P}, \qquad (1.4a)$$

where $\propto$ represents the proportionality symbol, or alternatively:

$$P \cdot V = constant. \qquad (1.4b)$$

Subsequent work by Charles (1746–1823) showed that at constant pressure (P) the volume (V) of a given amount of a gas is directly proportional to its absolute temperature (T). *Charles's law* states:

$$V \propto T. \qquad (1.5a)$$

Or alternatively,

$$\frac{V}{T} = constant. \qquad (1.5b)$$

Next, Avogrado (1776–1856) related volume (V) and the amount (n in moles), but at fixed temperature and pressure. *Avogrado's law* states:

$$V \propto n, \tag{1.6a}$$

Or alternatively,

$$\frac{V}{n} = constant. \tag{1.6b}$$

These three gas laws can be combined because volume (V) appears in all three:

$$V \propto \frac{nT}{P}. \tag{1.7}$$

The proportionality symbol ($\propto$) can be turned into an equality through the introduction of a proportionality constant (R):

$$V = R \cdot \frac{nT}{P}, \tag{1.8a}$$

This equation is usually written as:

$$PV = nRT, \tag{1.8b}$$

which is the well-known *ideal gas law*. R is the *universal gas law constant*, which has the value of 8.314 J mol^{-1} K^{-1} in SI units, when n is expressed in moles. Note that thermodynamic data are often provided in various units (e.g., pressure in bar, atm or tor rather than Pa; volume in liters rather than m^3); the units and values of R then need modification to maintain consistency of units. For instance, the value of R is 0.0820568 when expressed in L atm mol^{-1} K^{-1}.

1.3   System, Surrounding and Equations of State

Systems are a central concept in thermodynamics. The object of interest is defined as the system while everything else, the rest of the universe, is defined as surroundings. A system can be separated from the surroundings via imaginary boundaries or physically real boundaries such as the wall of a container. The system can be a chemical reaction taking place in a solution (e.g., calcite dissolution), it can be your cup of hot coffee cooling while you are enthusiastically reading this text, or the entire System Earth receiving short-wave radiation from the sun and emitting long-wave radiation to the surrounding cosmos.

A system interacts with its environment via mass transfer, or energy transfer via heat or work exchange (Fig. 1.2). An *isolated system* does not exchange heat, work, or matter with the surroundings. A *closed system* does not exchange matter but can exchange energy with the environment. An *open system* exchanges matter with the surroundings and may exchange energy (as heat or work) as well. Changes to the system are called *isothermal* if the temperature is kept constant, *isobaric* if pressure is kept constant, *isochoric* or isovolumetric if the volume is constant, and *adiabatic* if no heat is exchanged, i.e., if it occurs in thermal isolation (Fig. 1.3).

The state of a macroscopic system is described using observable quantities, i.e. state variables, such as pressure, temperature, volume, and number of moles. Equations of state relate the various state variables of a system. The ideal gas law (Eqs. 1.8a and 1.8b) is an equation of state for an ideal gas. For one mol of a gas ($n = 1$), the system is fully determined with two out of the three variables (V, T or P). For instance, for a standard temperature and pressure (STP, 0 °C and 1 bar) the gas would then

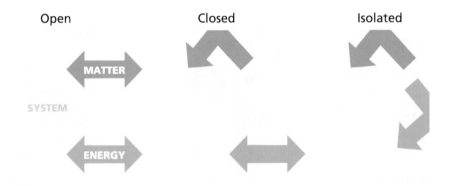

Fig. 1.2 Open systems exchange matter and energy; Closed systems exchange energy but no matter; Isolated system can exchange energy nor matter

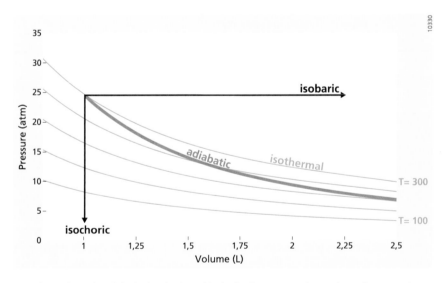

Fig. 1.3 Isothermal, adiabatic, isochoric and isobaric changes to a thermodynamic system (e.g., a gas)

have a volume of 22.4 L. For SATP conditions (25 °C and 1 bar) the molar volume (V_m) of an ideal gas would be about 24.5 L mol^{-1}.

Thermodynamic variables are referred to as *intensive* if they are independent of the amount, while *extensive variables* are additive, i.e., they depend on the sample size. Of the four variables in the gas law (Eqs. 1.8a and 1.8b), P and T are independent of the amount of gas and are referred to as intensive variables, while V and n are proportional to the amount of gas and are extensive variables. In thermodynamics, changes in extensive quantities are associated with changes in the respective specific intensive quantities and their product has the dimension of energy: e.g., PV in the ideal gas law. The ratio of two extensive variables can be an intensive variable, e.g., density (mass divided by volume).

The energy of a system is related to all the other measurables of the system via equations of state. Thermodynamics literally means 'heat movement' because it describes how the energy of a system relates to measurable variables. If the state of a system shows no tendency to change, i.e., the system is at equilibrium, the state functions have values which are independent of the history of the system. Consequently, the changes in a function of state do not depend on the route by which one goes from one state to another, i.e., they are path independent. Equilibrium thermodynamics, the topic of this course, focuses on differences in system states and we use the symbol Δ to indicate changes in state variables and functions.

Box 1 Math intermezzo: Partial derivatives and state functions

Equations of state usually involve multiple variables. The *total differential* of a function F (x,y,z) is defined as:

$$dF = \left(\frac{\partial F}{\partial x}\right)_{y,z} dx + \left(\frac{\partial F}{\partial y}\right)_{x,z} dy + \left(\frac{\partial F}{\partial z}\right)_{x,y} dz \quad (1.9)$$

where the derivative of F is taken with respect to one variable at a time while the others are kept constant. The first term $\left(\frac{\partial F}{\partial x}\right)_{y,z}$ is the derivative of F with respect to x only and is the *partial derivative*. These partial derivatives sometimes reveal relationships between state variables and are often used to formally define basic properties such as heat capacity, compressibility, etc.

As an example, suppose we aim to quantify the pressure dependence on temperature of an ideal gas, assuming that the volume and number of molecules remain constant. The relevant partial derivative is

$$\left(\frac{\partial P}{\partial T}\right)_{V,n}$$

We start with rewriting the ideal gas law (Eqs. 1.8a and 1.8b) to isolate pressure on one side of the equation:

$$P = \frac{nRT}{V}$$

Next, we take the derivative of both sides with respect to temperature T, while considering n and V constant, and obtain

$$\left(\frac{\partial P}{\partial T}\right)_{V,n} = \frac{\partial}{\partial T}\left(\frac{nRT}{V}\right) = \frac{nR}{V}\frac{\partial}{\partial T}T = \frac{nR}{V}. \qquad (1.10)$$

In this way, we have analytically derived how pressure varies with temperature and have revealed the phenomenological *law of Guy-Lussac* stating that, for a given mass (n) and volume (V), the pressure is proportional to absolute temperature, i.e., $\Delta P/\Delta T = $ constant.

Chapter 2
The First Law: Work, Heat and Thermochemistry

Abstract This chapter presents the first law of thermodynamics in terms of heat exchange and work. Internal energy, heat capacity and enthalpy are defined and enthalpy changes during reactions and phase changes are discussed. Latent and sensible heat are introduced, and adiabatic processes are related to the atmospheric lapse rate, potential temperature, and the geothermal gradient.

Keywords Zeroth Law · First Law · Heat · Work · Enthalpy · Heat capacity · Latent heat · Sensible heat · Adiabatic processes · Phase changes

Before elaborating on the first law and its implications, it is necessary to introduce the zeroth law of thermodynamics and to further define our systems. Consider two systems, A and B, that are closed, i.e., no matter can exchange, but with different temperatures. If these are brought in contact, heat will flow spontaneously from the hotter to the colder system till temperatures become equal, i.e., thermal equilibrium is reached. Energy transfer from one system to another due to temperature differences is called *heat*. If a third closed system, C, is in thermal equilibrium with system A, then $T_C = T_A$ and system C must be in thermal equilibrium with system B also. This illustrates the *zeroth law* of thermodynamics stating that

> if two systems are in thermal equilibrium with each other and a third system is in thermal equilibrium with one of them, then it is in thermal equilibrium with the other also.

Systems exchange matter and/or energy with their surroundings. Thermodynamics has adopted a system-centric view, i.e., matter added to, heat delivered to, and work done on the system have a positive sign. Conversely, work done by the system or heat flowing out of the system to the surroundings have a negative sign (Fig. 2.1).

J. J. Middelburg, *Thermodynamics and Equilibria in Earth System Sciences: An Introduction*, SpringerBriefs in Earth System Sciences,
https://doi.org/10.1007/978-3-031-53407-2_2

Fig. 2.1 A closed system
interacts with its
environment via work (w)
and heat (q) exchange. Heat
added to and work done on
the system are positive by
definition

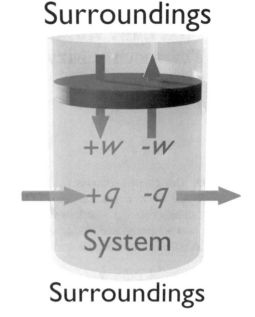

2.1 Work

Energy is defined as the capacity to do work. Work is being done when motion occurs
upon the action of a force. *Work*, w, is quantified as the product of force, F, and the
distance, d, over which it acts:

$$w = F \cdot d. \tag{2.1}$$

Work has the unit of Joule ($1\,J = 1\,Nm$), like other forms of energy. In thermodynamics a common form of work is expansion/compression work because of volume
changes of the system. Consider a cylinder with a piston (Fig. 2.2). Any process that
increases the gas volume within the cylinder under the piston will push the piston
upwards and in this way the system performs work on the surroundings, i.e., it loses
some energy in the form of work. The amount of work (w) done by the system for a
change in volume (ΔV) is

$$w = -P_{ext}\Delta V, \tag{2.2}$$

where P_{ext} is the external pressure of the surroundings, and the minus sign is needed
to satisfy the convention that work done by the system should be negative (Fig. 2.1).

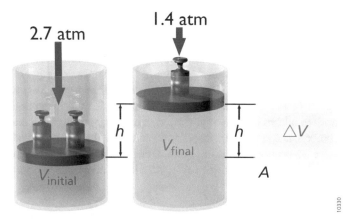

Fig. 2.2 Expansion work. Left: A piston exerts 2.7 atm on the gas within the cylinder and the gas has volume ($V_{initial}$). Middle/right: Following lowering of the external pressure to 1.4 atm, the volume of the gas increases ($\Delta V = V_{final} - V_{initial} =$ h(eight) × A(rea)) because of work done by the system $w = -P_{ext}\Delta V$

2.2 Heat

Heat, also called thermal energy, is denoted by the letter q and has the unit of joule. Traditionally, the calorie (cal) unit has been used for heat. One calorie is the energy needed to heat 1 mL of water from 15 to 16 °C, which is equal to 4.184 J. The amount of heat needed to change the temperature of a system depends on the temperature change, ΔT, and the heat capacity (C):

$$q = C \cdot T \tag{2.3}$$

The heat capacity C is an extensive property that depends on the amount of material in the system. The *molar heat capacity, C_m*, is the amount of heat needed to raise the temperature of 1 mol by 1 K and has the unit J mol^{-1} K^{-1}, while the *specific heat capacity, C_s*, is the amount of heat for 1 K temperature increase per gram of material with the unit J g^{-1} K^{-1}. Using the molar or specific heat capacities Equation (2.3) would then respectively become:

$$q = nC_m\Delta T, \tag{2.4}$$

and

$$q = mC_s\Delta T \tag{2.5}$$

where m is mass (g) and n is the number of moles. As an example, consider the energy required to heat 18 gram (1 mole) of water with a specific heat capacity (C_s of 4.18 J g^{-1} K^{-1}) from 25 to 50 °C. Using Eq. (2.5), we then arrive at 1882 J.

2.3 The First Law

The first law of thermodynamics can be formulated in multiple ways but in fact states: *energy can neither be created nor destroyed, i.e., energy is conserved.* The total energy of a system is defined as the *internal energy* for which we use the symbol *U*. In all phases (gas, liquid, solid) atoms and molecules are in motion. Internal energy comprises all types of energy: e.g., kinetic energy from molecular rotation and bond vibration, and potential energy from intermolecular attraction and chemical bonds. Because we cannot measure all types of energy in a system, the absolute total internal energy cannot be known. However, differences in internal energy (ΔU) can be measured and calculated.

An isolated system does not exchange matter or energy with the surroundings; hence the total internal energy of the system does not change. This leads to the statement of the *first law* of thermodynamics:

For an isolated system, the total energy of the system remains constant.

$$\text{Mathematically, this means: } \Delta U = 0 \tag{2.6}$$

This statement of the first law has limited utility because it is for an isolated system. An alternative, more useful way to express the first law (for a closed system) is the *change in internal energy is the sum of heat (q) added and work (w) done on the system*, mathematically:

$$\Delta U = q + w \tag{2.7}$$

This is the formulation of the first law used in this course. The internal energy *U* is a state function, i.e., it is path independent, while the heat *q* and work *w* are not state functions and depend on the path. In more advanced treatments, this equation is usually presented in calculus form as $dU = \delta q + \delta w$, where *d* represents an infinitesimally small change and δ is used to articulate that infinitesimally small changes in *q* and *w* depend on the path.

Another useful way of phrasing the first law is to consider expansion work ($w=-P\Delta V$):

$$\Delta U = q - P\Delta V \tag{2.8}$$

If we add heat to a system with a constant volume, e.g., a container, there is no P-V work, ΔV=0, then

$$\Delta U = q + 0 = q_V, \tag{2.9}$$

where q_V is the heat transfer at *constant volume*. Accordingly, the change in internal energy for a system at a constant volume is equal to the heat exchange that can be measured. This is very useful for engineers and others working with volume constrained systems (containers, reactors). However, for earth scientists studying processes at a certain pressure, or chemists studying reactions in the laboratory at atmospheric pressure, it is more useful to quantify the change in heat at a constant pressure, i.e., the **enthalpy change, ΔH**:

$$\Delta H = q_P = \Delta U + P\Delta V, \tag{2.10}$$

where q_P is the heat transfer at *constant pressure*. This equation directly follows from Eq. (2.8) and leads to the formal definition of enthalpy, H, as

$$H \equiv U + PV \tag{2.11}$$

where the symbol $\equiv$ is used for a formal definition.

Although the change in internal energy (ΔU) is the more fundamental quantity (used by physicists), the change in enthalpy (ΔH) is usually easier to measure and thus more commonly used. At constant P, the difference between ΔU and ΔH is due to the $P\Delta V$ term (Eq. 2.10). Volume changes involving liquid or solid phases are small, hence the $P\Delta V$ can be ignored and $\Delta H \approx \Delta U$. However, volume changes can be large if gases are involved and then enthalpy changes are a little higher than internal energy changes (≈ 2.48 kJ per mol at 25 °C). Enthalpy changes follow the sign convention introduced earlier (Fig. 2.1): processes releasing heat are *exothermic* and have a negative ΔH because the system loses energy, heat uptake by a system corresponds to positive ΔH values and are called *endothermic*.

The heat capacity *(C)* was defined as the amount of heat required to increase the temperature of a substance by 1K. Similar to heat (q), the heat capacity of a gas also depends on conditions, i.e., whether the process occurs under constant volume or constant pressure conditions. In case of constant volume, all the heat will increase the internal energy, and thus the temperature, because no expansion work is done. At a constant pressure, some of the heat goes into work and thus more energy is needed to cause the same temperature rise. The *constant pressure heat capacity* C_P is thus higher than the *constant volume heat capacity* C_V. Mathematically, C_V relates ΔT to the change in internal energy:

$$\Delta U = q_V = C_V \Delta T \tag{2.12}$$

whereas C_P links ΔT to the enthalpy change

$$\Delta H = q_P = C_P \Delta T. \tag{2.13}$$

Volume change with heating of liquids and solids is generally limited and their C_V and C_P are very close in value and any difference can usually be neglected. For an ideal gas, *Mayer's relation* links C_V and C_P:

$$C_P = C_V + nR \tag{2.14}$$

Having discussed the internal energy changes of a gas at a constant temperature and constant pressure, we now consider an isothermal change. Isothermal conditions imply that ΔT is zero, hence $\Delta U = 0$ (Eq. 2.12). The first law then reads

$$\Delta U = 0 = q + w$$

and consequently, for *isothermal processes*

$$q = -w \tag{2.15}$$

An isothermal compression (decrease in volume) causes a net outflow of heat, while isothermal expansion (increase in volume) must be compensated by heat uptake to maintain constant temperature.

2.4 Thermochemistry: Enthalpy Changes in Chemical Reactions

All chemical substances have enthalpy stored in the form of chemical bonds. When a chemical reaction occurs, the change in enthalpy, ΔH, is equal to the total enthalpy of products (final condition with new chemical bonds) minus the total enthalpy from the reactants (initial condition with pre-existing chemical bonds)

$$\Delta_r H = H_{products} - H_{reactants}, \tag{2.16}$$

where $\Delta_r H$ is the enthalpy change of a reaction, as indicated by the subscript r. However, absolute enthalpies of substance cannot be determined (like internal energy), only relative values, i.e., changes in enthalpy can be determined. In other words, we need a set of standards against which enthalpies of reaction can be measured. The *standard enthalpy of formation of a certain compound*, ΔH_f^o, is defined as the *enthalpy change for the formation of one mole of that compound from its elements in the most stable form under standard conditions* (STAP, 25 °C and 1 bar). The subscript $_f$ is added to indicate that is the enthalpy of formation and the superscript ^{o} that it relates to pure substances at SATP. For example, the gas N_2 and the solid graphite are the most stable form of elements N and C at standard conditions and their enthalpies of formation (ΔH_f^o) have been set to zero. Thermochemists have constructed consistent databases with standard enthalpies of formation that can be used to calculate the enthalpy of reaction at standard conditions and making use of reaction stoichiometry.

Specifically, for the reaction of α moles of substance A and β moles of substance B to form γ moles of substance C and δ moles of substance D

$$\alpha A + \beta B \rightarrow \gamma C + \delta D \tag{2.17}$$

the enthalpy of reaction at SATP ($\Delta_r H^o$) is calculated as follows,

$$\Delta_r H^o = \left[\gamma \times \Delta H_f^o(C) + \delta \times \Delta H_f^o(D) \right] - \left[\alpha \times \Delta H_f^o(A) + \beta \times \Delta H_f^o(B) \right]$$
(2.18)

or more generalized:

$$\Delta H_r^o = \sum v_i \Delta H_f^o(products) - \sum v_i \Delta H_f^o(reactants),$$
(2.19)

where v_i is the stoichiometric coefficient of substance i in the reaction.

To illustrate this method to calculate $\Delta_r H^o$, consider the oxidation of 1 mole of sucrose ($C_{12}H_{22}O_{11}$; s) to CO_2 (aq) and water (l) at 1 bar. The relevant reaction and enthalpy of formations are:

Reaction	$C_{12}H_{22}O_{11}(s) +$	$12O_2(g)$	$\rightarrow$	$12CO_2(aq)+$	$11H_2O(l)$
ΔH_f^o	-2226.1 kJ mole^{-1}	0 kJ mole^{-1}		-412.9 kJ mole^{-1}	-258.8 kJ mole^{-1}

and the enthalpy of reaction is calculated using 2.18:

$$\Delta H_r^o = \left[12\Delta H_f^o(CO_2(aq)) + 11\Delta H_f^o(H_2O(l)) \right]$$
$$- \left[\Delta H_f^o(C_{12}H_{22}O_{11}(s)) + 12 \times \Delta H_f^o(O_2(g)) \right]$$
$$= 12 \times -412.9 + 11 \times -285.8 - 2226.1 - 12 \times 0 = -5873 \text{ kJ}$$

This enthalpy of reaction is also known as *enthalpy of combustion* or *heat of combustion*, i.e., the enthalpy of change when one mole of a compound reacts completely with excess oxygen gas. When using thermodynamic data, one should note that each compound and each phase of a certain compound have distinct internal energies and enthalpies and that the phase [*solid (s), liquid (l), dissolved (aq) and gas (g)*] should therefore explicitly be indicated to use the correct energy. This is because, as we will see below, phase transitions cause enthalpy changes.

This is one way to obtain enthalpy of reactions, the other method is based on *Hess's law*.

Since enthalpy is a state function, only the initial and final conditions do matter, and the pathway (intermediate reactions) does not matter. Hess's law states that *the overall enthalpy change for a reaction is equal to the sum of enthalpy changes for the individual steps in the reaction.*

To illustrate Hess's law, consider the Haber(-Bosch) process, the common way to produce ammonia (needed for fertilizers to produce food for the world population) in large quantities from nitrogen and hydrogen gas. It is a multiple-step process:

Step 1 : $2H_2(g) + N_2(g) \rightarrow N_2H_4(g)$ $\Delta H_1 = ?$

Step 2 : $N_2H_4(g) + H_2(g) \rightarrow 2NH_3(g)$ $\Delta H_2 = -187.6$ kJ

Combined : $3H_2(g) + N_2(g) \rightarrow 2NH_3(g)$ $\Delta H_{1+2} = -92.2$ kJ

(net reaction)

Hess's law implies that $\Delta H_1 + \Delta H_2 = \Delta H_{1+2}$; hence $\Delta H_1 = \Delta H_{1+2} - \Delta H_2 = +95.4\,kJ$. The first step is thus endothermic consuming heat (the system/reaction receives heat), while the overall reaction is exothermic and produces heat.

The heat of combustion or enthalpy of combustion is usually presented in $kJ\,g^{-1}$ (i.e. *specific combustion enthalpy*) rather than $kJ\,mol^{-1}$, the unit used for enthalpy of formation $\left(\Delta H_f^o\right)$. This specific combustion enthalpy is a measure of the energy density.

While the heats of combustion of methane, ethane, propane, butane and pentane are -890, -1560, -2220, -2877 and $-3509\,kJ\,mol^{-1}$ and differ substantially, their respective specific combustion enthalpies are more alike: -55.5, -51.9, -50.4, -49.5 and $-48.6\,kJ\,g^{-1}$, respectively. For comparison, typical specific combustion enthalpies in $kJ\,g^{-1}$ of kerosene (-46.2), diesel (-44.2), crude oil (-43) are also in this range, while those of anthracite coal (-32.5), lignite coal (-15), wood (-15 to -20) and peat (-15 to -20) are much lower.

2.5 Enthalpy Changes During Phase Changes

Matter exists in three phases (solid, liquid and gas) and transformation from one phase to another involves heat transfer, i.e., a change in enthalpy (Fig. 2.3). The transition from solid to liquid, melting or fusion, requires addition of heat (endothermic process). The *enthalpy of fusion* ΔH_{fusion} is defined as the amount of heat necessary to melt a substance at its melting point without changing its temperature at constant pressure. Standard enthalpies of fusion ΔH_{fusion}^o are available at 1 bar and 25 °C. Freezing or solidification, the opposite process of melting, is exothermic and the enthalpy change has the same magnitude as that of fusion but the opposite sign (negative for freezing, positive for fusion).

Vaporization (or boiling) is the transformation of a liquid into gas or vapor and the *enthalpy of vaporization* ΔH_{vap} is the energy required at constant pressure to vaporize one mole of pure liquid at its boiling temperature. Condensation, the opposite of vaporization, has the same enthalpy as vaporization but the opposite sign (negative for condensation, positive for vaporization).

Sublimation is the transformation of a solid into a gas. The *enthalpy of sublimation* ΔH_{subl} is the amount of heat required to convert a solid into a gas without going through the liquid phase. An example is dry ice (solid carbon dioxide) which at atmospheric pressure goes directly into a gas without formation of liquid carbon dioxide at a temperature of -78.5 °C. Condensation or deposition, the reverse process has again a different sign for the enthalpy change than sublimation. At a constant temperature, enthalpies of sublimation can be directly calculated from the enthalpies of fusion and vaporization using Hess's law:

$$\Delta H_{subl} = \Delta H_{fusion} + \Delta H_{vap} \qquad (2.20)$$

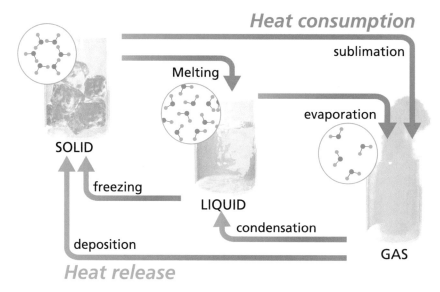

Fig. 2.3 Phase transformations involving solid, liquid and gas. Melting, sublimation and evaporation involve heat consumption, while freezing, deposition and condensation release heat

2.6 Turning Ice into Steam: Latent and Sensible Heat

Water plays a major role in System Earth: it dominates the cryosphere (and obviously the hydrosphere) and impacts the biosphere, geosphere and atmosphere. We have covered enough material to calculate the temperature response during the heating of 1 gram of ice with an initial temperature of −20 °C to a steam of 120 °C at earth surface pressure of 1 atm (Fig. 2.4).

Upon addition of heat to ice of −20 °C, the ice will heat up and with a specific heat capacity (C_s) of 2.09 J g^{-1} K^{-1}. For one gram of ice, 41.8 J (20 × 2.09) will be required to reach the melting point of ice. At the melting point, 0 °C, further addition of heat will melt the ice but the system remains at the same temperature, and 331 J will be needed to convert one gram of ice into liquid (enthalpy of fusion). Further addition of heat will warm up the water and 418 J are required to reach the boiling point at 100 °C (since water has a specific heat capacity of 4.18 J g^{-1} K^{-1}). At the boiling point, liquid water will gradually turn into steam (evaporate), and this requires 2260 J (enthalpy of vaporization), but temperature does not change. Finally, further addition of heat will increase the temperature of the steam. To reach 120 °C, this will require an additional 36.8 J, because steam has a specific heat content of 1.84 J g^{-1} K^{-1}. The total energy required is about 3088 J, of which 84 % is used for phase changes (10.7 % for melting and 73.2 % for boiling) and does not result in a temperature increase. The energy consumed or released during phase transitions at a constant temperature is also known as 'hidden' or *latent heat*. This contrast with *sensible heat,* that results in a temperature change of the system. Latent

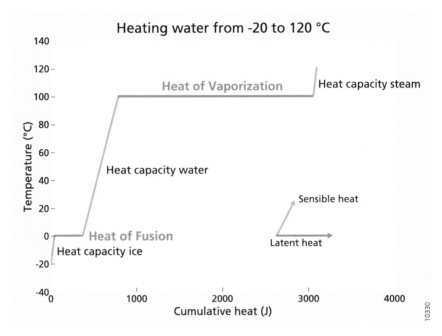

Fig. 2.4 Latent and sensible heat during warming of ice to steam

heat fluxes are an important component of Earth's surface energy budget, e.g., the evaporation/transpiration at the Earth surface and subsequent condensation of water in the troposphere. In the Earth's interior, latent heat is also released if the liquid outer core crystallizes at the inner core boundary.

2.7 Adiabatic Changes, Lapse Rate, Geothermal Gradient and Potential Temperature

Adiabatic processes are thermodynamic state changes without heat transfer ($q = 0$). Accordingly, the first law $\Delta U = q + w$ then reduces to

$$\Delta U = w = -P\Delta V \tag{2.21}$$

stating that the change in internal energy of a system equals work. Adiabatic processes are usually associated with a temperature change of the system because of changes in volume. Compression (due to increasing pressure) causes temperature increases and expansion results in cooling due to reducing pressure. If a system expands or compresses very fast, or heat exchange occurs very slowly, it will not have time to exchange heat with its surroundings. Describing the system volume change as an adiabatic process is then an appropriate approximation. For instance, air transport upward in the atmosphere can be modelled as an adiabatic process because the heat transfer between the air parcel and the rest of the atmosphere is slow on the timescale of its upward motion. A rising magma plume or air escaping after opening a bottle of champagne can also be treated as an adiabatic process. The geothermal gradient can be explained by adiabatic compression (Box 2).

For an adiabatic process, the change in internal energy can be determined from two expressions (Eqs. 2.21 and 2.12):

$$\Delta U = -P\Delta V \text{ and } \Delta U = C_V \Delta T$$

In which C_V is the heat capacity at constant volume.

Equating these two expressions

$$C_V \Delta T = -P\Delta V \tag{2.22}$$

shows that compression ($\Delta V < 0$) leads to a positive ΔT, hence an increase in temperature, while expansion ($\Delta V > 0$) results in a negative ΔT and is thus accompanied by a decrease in temperature.

Adiabatic expansion can quantitatively explain the atmospheric *lapse rate*, i.e., the well-known decrease in temperature while moving upward in the atmosphere. Mathematically, lapse rate (Γ) is the temperature gradient:

$$\Gamma = -\frac{\Delta T}{\Delta z}, \tag{2.23}$$

where z is height (in m).

To link the lapse rate with the first law, we treated an air parcel as a closed system and differentiate the ideal gas law for one mol ($PV = RT$):

$$P\Delta V + V\Delta P = R\Delta T$$

and then re-arrange to isolate ΔT

$$\Delta T = \frac{P\Delta V + V\Delta P}{R}$$

Then substitute it into Eq. (2.22):

$$C_v\left(\frac{P\Delta V + V\Delta P}{R}\right) + P\Delta V = 0 \tag{2.24a}$$

using Mayer's relation (Eq. 2.14) $R = C_P - C_V$, we obtain

$$C_v \left(\frac{P \Delta V + V \Delta P}{C_P - C_V} \right) + P \Delta V = 0 \qquad (2.24\text{b})$$

Which can be rewritten as

$$P \Delta V = -\frac{V \Delta P}{\frac{C_P}{C_V}}. \qquad (2.25)$$

This is the thermodynamic equation for adiabatic processes. It is sometimes presented in integrated form as $P V^{\frac{C_P}{C_V}} = constant$. Substituting Eq. (2.25) into the first law for adiabatic processes (Eq. 2.22) using the specific heat capacity at constant volume, we obtain

$$m C_V \Delta T = \frac{V \Delta P}{\frac{C_P}{C_V}}$$

which can be rewritten as

$$\rho C_P \Delta T = \Delta P \qquad (2.26)$$

where ρ is $\frac{m}{V}$, the density of the gas.

Combining this equation with the equation for hydrostatic equilibrium of the atmosphere,

$$\Delta P = -\rho g \Delta z \qquad (2.27)$$

with g as the standard gravitation acceleration at the earth surface (9.8 m s^{-2}) and z as height (m) to eliminate the pressure term, we eventually arrive at the lapse rate:

$$\Gamma = -\frac{\Delta T}{\Delta z} = \frac{g}{C_P} = 9.8 \text{ K km}^{-1}$$

since the specific heat capacity C_P value for dry air is about $1 \text{ kJ kg}^{-1} \text{ K}^{-1}$. This theoretical estimate for the dry air is fully consistent with the experience-based lapse rate of $1 \,°C$ per 100 m (Fig. 2.5).

However, air contains moisture that condenses upon cooling (cloud formation) and then releases (latent) heat (ΔH). This heat is added to the internal energy of the air. In this case we must extend Eq. (2.22) with latent heat:

$$m C_V \Delta T = -P \Delta V + \Delta H \qquad (2.28)$$

And since ΔH is positive (added to air), the adiabatic cooling (ΔT) will be less, thus the environmental lapse rate will be smaller, consistent with observations (Fig. 2.5). Moist adiabatic lapse rates are typically $0.5 \,°C$ per 100 m.

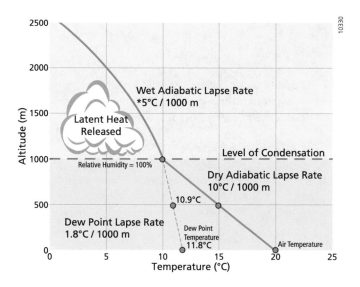

Fig. 2.5 Dry adiabatic lapse rate applies till level of condensation, above which latent heat is released (Eq. 2.28) and the lapse rate becomes less

Pressure change induced adiabatic processes increase temperatures upon compression and decrease temperatures following expansion. Atmospheric and ocean scientists therefore use *potential temperature*, which is defined as *the temperature that a parcel of air/water would attain if adiabatically brought to a standard reference pressure*, i.e., earth surface conditions. Meteorologists use a standard pressure of 1000 mbar (1000 hPa) for reference, while oceanographers use the ocean surface pressure (0 dbar). Changes in temperature solely caused by compression or expansion are generally not of interest when studying atmosphere or ocean dynamics or heat content.

The potential temperature (θ) of an ideal gas is given by

$$\theta = T \left(\frac{P_0}{P} \right)^{\frac{R}{C_P}} \tag{2.29}$$

where P is pressure, P_0 is the reference pressure, T is the current temperature in K, R is the gas constant and C_P is the specific heat capacity at constant pressure. Meteorologists use a value 0.286 for the ratio $\frac{R}{C_P}$. Potential temperature (θ) in the ocean is always lower than the actual temperature by about 0.1 K for every km depth increase. This limited increase of in situ water temperature relative to potential temperature is due to the large heat capacity of water.

Box 2 Geothermal gradient of the inner Earth

The geothermal gradient, i.e., the temperature increase with depth into the Earth, can also be explained by adiabatic compression. The temperatures of the Earth's mantle and outer core are close to the adiabatic temperature curve. For adiabatic compression, a similar analysis as for the atmospheric lapse rate due to expansion can adopted, starting with a modified form of Eq. (2.26)

$$mC_P\Delta T = TV\alpha_P\Delta P \tag{2.30}$$

where m is mass, C_P is the specific heat capacity of the minerals, and α_P is the fractional increase in volume per degree increase in temperature. Specifically, the volume change in atmospheric adiabatic processes $\frac{V}{\frac{C_P}{C_V}}$ in Eq. (2.25) relates to the volume change $TV\alpha_P$ in the inner Earth adiabat (Eq. 2.30).

Re-arranging this equation to isolate the adiabatic change of temperature with pressure:

$$\frac{\Delta T}{\Delta P} = \frac{TV\alpha_P}{mC_P} = \frac{T\alpha_P}{\rho C_P} \tag{2.31}$$

where ρ is the density $\frac{m}{V}$.

Using the appropriate hydrostatic pressure Eq. (2.27), we obtain after re-arrangement

$$\frac{\Delta T}{\Delta z} = \frac{gT\alpha_P}{C_P}. \tag{2.32}$$

Using representative values for physical properties in the Earth's interior (Table 2.1), we can calculate the adiabatic temperature gradients: $\frac{\Delta T}{\Delta z} = 0.88$ Kkm^{-1} at the core-mantle boundary and $\frac{\Delta T}{\Delta z} = 0.29$ Kkm^{-1} at the inner-core boundary.

Table 2.1 Physical parameters in the outer and inner core near to the core-mantle boundary (CMB) and inner-core boundary 9ICB) (Lowry, 2011)

Property	Unit	Core-mantle boundary	Inner core boundary
Gravity, g	m s^{-2}	10.7	4.4
Density, ρ	kg m^{-3}	9900	12,980
C_P	J K^{-1} kg^{-1}	815	728
T	K	3700	5000
Volume expansion coefficient, α_P	10^{-6} J kg^{-1}	18.0	9.7

Chapter 3
Entropy and the Second Law

Abstract This chapter presents the second and third laws of thermodynamics, both at the macroscopic and microscopic scale. The state variable entropy is introduced, and it changes during reactions are quantified. A spontaneous process is recognized to cause an increase in entropy.

Keyword Second Law · Third Law · Entropy · Spontaneous process

The first law has shown to provide much insight into how nature functions, but it does not explain why some processes occur spontaneously. A *spontaneous* process proceeds in one direction, and once initiated, will continue. For instance, if you drop a rock from some table height, it will fall spontaneously, but a rock on the ground does not jump back to table height spontaneously. If we have two flasks, one at vacuum and the other one at 1 bar, and connect these, the pressure will equalize to 0.5 bar in each of the flasks, but without outside intervention gases will never unmix so that one flask will attain vacuum and the other 1 bar again. Heat will always flow from a warm to cold reservoir and not the other way around. These spontaneous processes cause a decrease in useable energy: gravitational potential energy for falling rock, work for the expanding gas and thermal energy for the heat flow. A process such as the dissolution of sodium chloride in water is spontaneous, yet it is an endothermic process ($\Delta H_r^o = +3.88$ kJ mol^{-1}), consuming enthalpy and cooling the water. Apparently, changes in energy constrained by the first law are insufficient to explain the spontaneous dissolution of kitchen salt while you are cooking. Another thermodynamic law (the second) and a property called entropy are required. *Entropy* is a thermodynamic state function denoted by the symbol S, which is explained below.

© The Author(s) 2024

J. J. Middelburg, *Thermodynamics and Equilibria in Earth System Sciences: An Introduction*, SpringerBriefs in Earth System Sciences, https://doi.org/10.1007/978-3-031-53407-2_3

3.1   The Second Law

There are multiple formulations of the second law that emphasize its various implications. The second law not only makes a statement of the direction of change, but also
puts constraints on the efficiency of energy conservations. The latter aspects of the
second law are often illustrated with a Carnot cycle of a heat engine and are primarily
the domain of physicists and engineers, while chemists give more attention to the
direction of reactions. While work can be converted to heat with 100% efficiency, this
is not the case for the conversion of heat to work. There is an inherent loss of useable
heat when it is used to perform work. This observation underlies *Clausius'* statement
of the 2nd law *that heat can never pass from a colder to a warmer body without some
other change, connected therewith, occurring at the same time.* Another formulation
for the 2nd law by *Kelvin* states *there exists no process in which the sole result is the
complete conversion of heat absorbed by a system into work.* Clausius' statement
articulates the direction of spontaneous processes, while Kelvin's statement empha-
sizes the degradation of energy when it is transferred as heat (not all heat is available
for work). While the first law states that the *sum* of heat and work (=internal energy)
is conserved, the second law adds that the interconvertibility of heat and work is
asymmetrical. Another, useable formulation of the **second law** states that **spontaneous processes are those which increase the entropy of the universe**. Accordingly,
we must properly define the entropy change. Entropy will first be illustrated using
statistical thermodynamics (and on the way we will introduce the third law of thermodynamics) and then the macroscopic entropy definition will be presented and
elaborated.

3.2   Microscopic View of Entropy and the Third Law

The microscopic definition is based on statistical mechanics. Each macroscopic thermodynamic state has a specific *number of microstates*, W, associated with it. A
microstate is a particular way in which the total energy of the system is distributed
among the individual molecules of the system. Consider a gas distributed over two
bulbs, one is empty, the other one filled with four molecules (Fig. 3.1). After opening
the valves between these bulbs, the molecules are re-distributed and 16 possible
molecular arrangements can be found ($=2^4$; 2 bulbs and 4 four molecules), each
configuration is equally likely. The probability of the system returning to its original
state is $1/16$ ($=2^{-4}$). Now consider the case of one mole of gas in one of the bulbs and
the other bulb at vacuum again. One mole of gas contains 6.022×10^{23} molecules,
i.e., Avogrado's number (N_A). It will be clear that the probability for one mole of
gas containing $\approx 10^{24}$ molecules to return to its original state in one of the bulbs only
is extremely low ($\approx 2^{-10^{24}}$). Accordingly, the number of molecular arrangements, or
microstates W, increases during spontaneous changes. In other words, the randomness or disorder of the system increases in spontaneous processes. *Boltzmann* defined
the entropy as follows:

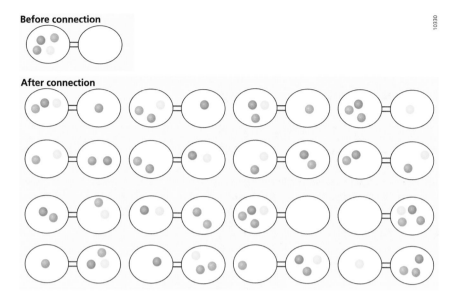

Fig. 3.1 Possible distribution of four molecules before and after connection of two gas bulbs

$$S = k_B \ln W. \qquad (3.1)$$

This equation links entropy (S) with the Boltzmann constant ($k_B = 1.381 \times 10^{-23}$ J K^{-1}) and the number of microstates (W). The Boltzmann constant is via Avogrado's constant (N_A) related to the macroscopic gas constant (R):

$$R = k_B \cdot N_A$$

This statistical thermodynamic view of entropy as a measure of randomness or disorder is instructive to explain some of its properties (e.g., its dependence on phase and temperature). Gas has more molecular randomness than liquids (because of the larger distance between molecules) and its entropy is therefore higher than that of liquids. Liquids allow movement of individual molecules and have therefore more randomness than solids, in which molecules are in fixed positions. Consequently, the entropy will decline when liquids freeze to form a solid phase. The dissolution of a well-organized sodium chloride crystal in water leads to increase in randomness because of the disruption of the crystal, but the hydration of the chloride and sodium ions somewhat lowers the randomness; the net result is that randomness and thus entropy increases, driving the dissolution of kitchen salt in water.

Lowering of temperature decreases the kinetic energy of molecules and atoms. When the absolute temperature of a system approaches 0 K, then the motion of individuals molecules approaches zero and there is only one microstate ($W = 1$). Accordingly, application of Boltzmann's law (Eq. 3.1) then indicates that the entropy should be zero. This constraint is known as the *third law of thermodynamics* stating that *the entropy of a perfect crystal at zero kelvin is zero.*

3.3 Macroscopic View of Entropy

The macroscopic definition is based on a detailed analysis of heat-work cycles and the efficiency of heat engines (Carnot cycle). *Carnot* has shown that cyclic processes involving isothermal and adiabatic changes of an ideal gas cause no change in (internal) energy because of the return to initial conditions. Consider a heat engine with two reservoirs at temperature T_h (the source) and T_c (the sink), receiving heat q_h and transferring heat q_c, respectively (Fig. 3.2). Because T_h is larger than T_c we can extract work (w) from the heat engine.

Carnot showed that the maximum efficiency (e) of such a heat engine with fixed reservoir temperatures is

$$e = \frac{w}{q_h} = \frac{T_h - T_c}{T_h} \tag{3.2}$$

The efficiency of an engine can thus only be 100% if T_c is absolute zero. The Carnot, i.e., ideal, efficiency of an automobile engine having a T_h of 1100 K and an

Fig. 3.2 A hot reservoir at T_h supplies 20 kJ heat q_h to a heat engine performing 5 kJ work (w) and the remaining 15 kJ is lost as heat (q_c) to the cold reservoir at T_c. The efficiency is the work done divided by the heat supplied (q_h)

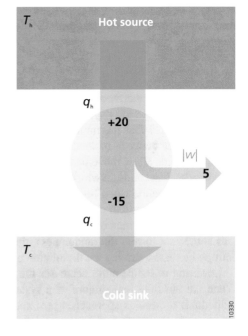

ambient temperature T_c of 295 K is about 73%; however, the average automobile engine is typically much less efficient due to other causes.

Conservation of energy implies that

$$w + q_h + q_c = 0. \tag{3.3}$$

Isolating w from this equation and inserting the expression in (3.2) then yields

$$\frac{q_c}{q_h} = -\frac{T_c}{T_h}, \tag{3.4a}$$

or its equivalent

$$\frac{q_h}{T_h} + \frac{q_c}{T_c} = 0. \tag{3.4b}$$

Which implies that the ratio $\frac{q}{T}$ is a state function whose change is zero in the Carnot cycle.

Following Clausius, the entropy change can then be defined as

$$\Delta S = \frac{q}{T}. \tag{3.5}$$

Next, we consider the spontaneous transfer of heat q from a hot reservoir (T_h) to a cold reservoir (T_c). The entropy decrease for the hot reservoir $\left(\frac{q}{T_h}\right)$ is smaller than the entropy increase of the cold reservoir $\left(\frac{q}{T_c}\right)$, and thus the overall entropy change ΔS is positive, as required by the second law (Fig. 3.3). Clausius inequality statement implies that for any spontaneous process

$$\Delta S > 0. \tag{3.6}$$

Fig. 3.3 Heat (q) spontaneously flows from a hot reservoir with temperature T_h to a cold reservoir at temperature T_c. The entropy change ΔS is positive because the entropy decrease for the hot reservoir $\left(\frac{q}{T_h}\right)$ is smaller than the entropy increase of the cold reservoir $\left(\frac{q}{T_c}\right)$.

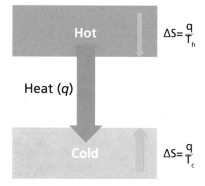

3.4   Entropy Changes During Reactions

Since entropy is a state function, the change in entropy ($\Delta_r S$) is the difference between the entropy of products and reactants:

$$\Delta_r S = S_{products} - S_{reactants}$$

Following the same approach as for enthalpy changes, we can obtain the entropy change ($\Delta_r S^0$) for the reaction of α moles of substance A and β moles of substance B to form γ moles of substance C and δ moles of substance D

$$\alpha A + \beta B \rightarrow \gamma C + \delta D$$

from the *standard molar entropies at STAP (S^0)* as follows

$$\Delta_r S^0 = \left[\gamma \times S^0(C) + \delta \times S^0(D)\right] - \left[\alpha \times S^0(A) + \beta \times \Delta S^0(B)\right] \qquad (3.7)$$

or more generalized:

$$\Delta_r S^0 = \sum v_i \Delta S^0(products) - \sum v_i \Delta S^0(reactants) \qquad (3.8)$$

The *standard molar entropy, S^0*, of a compound is *the molar entropy of a substance in its standard state at 25 °C and 1 bar pressure*. These values are tabulated at STAP conditions and normally expressed in $J\,K^{-1}\,mol^{-1}$ (rather than $kJ\,mol^{-1}$ as enthalpy). Moreover, these are absolute values because the third law sets S^0 at zero at zero K.

Consider the following reaction (Table 3.1):

$$2H_2(g) + O_2(g) \rightarrow 2H_2O(l) \qquad (3.9)$$

The entropy change of the system (=reaction) is calculated,

$$\Delta_r S^0 = \left[2 \times S^0(H_2O(l))\right] - \left[2 \times S^0(H_2(g)) + 1 \times \Delta S^0(O_2(g))\right]$$
$$= -326.68\,J\,K^{-1}$$

Table 3.1 Thermodynamic data for hydrogen and oxygen gas and liquid water

	$H_2(g)$	$O_2(g)$	H_2O (l)
ΔH_f^o, kJ mol^{-1}	0	0	−285.83
S^0, J K^{-1} mol^{-1}	130.68	205.14	69.91
ΔG_f^o, kJ mol^{-1}	0	0	−237.13

There is a loss of entropy because three moles of gas (disordered) are turned into two moles of liquid water (less molecules and less disordered). Taking at face value, this negative entropy change of the system (ΔS_{Sys}) suggests that water should not be stable and that the reverse reaction, transforming liquid water into hydrogen and oxygen gas, should be spontaneous. This is incorrect. The issue is that the second law states that spontaneous reactions should increase the *total entropy of the universe*, not that of the system only. To resolve this, we calculate the entropy change of the surroundings (ΔS_{Surr}) realizing that the enthalpy change of the surrounding ΔH^o_{surr} should be the negative of that of the system ($\Delta_r H^0 = \Delta H^o_{Sys}$):

$$\Delta H^o_{surr} = -\Delta H^o_{Sys} \qquad (3.10)$$

Using the above thermodynamic data and Eq. (2.18),

$$\Delta_r H^0 = 2(-285.83) - (2 \times 0 + 1 \times 0) = -571.66\,\text{kJ}$$

For constant P, we can use Eq. (3.5) (the entropy definition):

$$\Delta S^0_{surr} = \frac{q}{T} = \frac{\Delta H^o_{surr}}{T} = -\frac{\Delta H^o_{Sys}}{T} = \frac{571.66 \times 1000}{298} = 1918.3\ \text{J K}^{-1}$$

The total entropy change (ΔS_{Tot}) is thus.

$$\Delta S^0_{Tot} = \Delta S^0_{Sys} + \Delta S^0_{Surr} = -326.68 + 1918.3 = +1591.6\ \text{J K}^{-1}$$

In conclusion, for a spontaneous process

$$\Delta S_{Tot} = \Delta S_{Sys} + \Delta S_{Surr} > 0, \qquad (3.11)$$

Consistent with the 2nd law statement at the end of Sect. 3.1: *spontaneous processes are those which increase the entropy of the universe.* The entropy change of the system may be negative, but the total entropy change should be positive for any spontaneous process. This leads to a few simple statements:

$$\Delta S_{Tot} > 0 \text{ for a spontaneous process} \qquad (3.12a)$$

$$\Delta S_{Tot} < 0 \text{ for a non-spontaneous process} \qquad (3.12b)$$

$$\Delta S_{Tot} = 0 \text{ for an equilibrium} \qquad (3.12c)$$

Finally, any process taking place in the universe is spontaneous and leads to an increase of S_{Tot}. Consequently, the $\Delta S_{Tot} > 0$ criterion implies a unique direction of time and underlies the maximum entropy production concept in use in Earth System and Ecosystem sciences.

Chapter 4
The Gibbs Free Energy

Abstract This chapter introduces the Gibbs free energy and its relevance for the direction of phase changes and reactions. The dependence of Gibbs free energy on temperature and pressure is presented and used to derive the (Clausius-)Clapeyron relation. One-component phase diagrams and Gibb's phase rule are presented.

Keywords Gibbs free energy · Clapeyron relation · Gibbs' phase rule · Phase diagram

The second law provides a robust framework for predicting whether processes will be spontaneous or not. However, the assessment is made at the level of the universe rather than the system alone. To resolve this, the Gibbs free energy, another state function, must be introduced.

Starting with Eq. 3.11:

$$\Delta S_{Tot} = \Delta S_{Sys} + \Delta S_{Surr} > 0$$

we can replace ΔS_{Surr} using the definition of entropy (Eq. 3.5):

$$\Delta S_{Surr} = \frac{q_{surr}}{T} = \frac{\Delta H_{Surr}}{T}$$

Noting that the latter equality only applies under constant pressure conditions (see Sect. 2.3). The enthalpy change of the surroundings is the reverse of that of the system (Eq. 3.10): $\Delta H_{Surr} = -\Delta H_{Sys}$. Hence, we can obtain an equation for the system only:

$$\Delta S_{Sys} - \frac{\Delta H_{Sys}}{T} > 0$$

which after re-arrangement reads

$$\Delta H_{Sys} - T\Delta S_{Sys} < 0$$

© The Author(s) 2024

J. J. Middelburg, *Thermodynamics and Equilibria in Earth System Sciences: An Introduction*, SpringerBriefs in Earth System Sciences, https://doi.org/10.1007/978-3-031-53407-2_4

Table 4.1 Temperature and spontaneity of processes

ΔH	ΔS	$-T\Delta S$	ΔG	High T	Low T
−	+	−	−	Spontaneous	Spontaneous
−	−	+	±	Nonspontaneous	Spontaneous
+	+	−	±	Spontaneous	Nonspontaneous
+	−	+	+	Nonspontaneous	Nonspontaneous

and the subscripts can be dropped to eventually obtain an equation for spontaneous reaction based *on systems properties only*

$$\Delta H - T\Delta S < 0. \tag{4.1}$$

The *Gibbs (free) energy (G)* is formally defined as

$$G \equiv H - TS \tag{4.2}$$

and is a state function because the constituent terms (H, T, S) are state variables. The change in Gibbs free energy (ΔG) can thus be evaluated based on the difference between final and initial states.

Accordingly at constant temperature and pressure,

$$\Delta G = \Delta H - T\Delta S. \tag{4.3}$$

This useful equation provides the basis for predicting spontaneity, hence the direction of processes, a criterion for equilibrium and quantifies the energy available for work or metabolism of organisms. Comparing Eq. 4.3 with Eq. 3.12, it is clear that at constant P and T,

$$\Delta G < 0 \text{ for a spontaneous process} \tag{4.4a}$$

$$\Delta G > 0 \text{ for a non-spontaneous process} \tag{4.4b}$$

$$\Delta G = 0 \text{ for equilibrium, no tendency of the system to change} \tag{4.4c}$$

The Gibbs free energy function also provides *first-order guidance for the temperature dependence of chemical and phase equilibria*. The change in Gibbs free energy $\Delta G = \Delta H - T\Delta S$ contains two terms: a change in enthalpy or heat of reaction (ΔH) and the product of temperature and the change in entropy $(T\Delta S)$. The temperature dependence of the direction of phase transformations and chemical reactions relate to this second term. Processes can be divided into four categories based on the temperature response (Table 4.1).

The temperature at which processes switch from being spontaneous to becoming non-spontaneous can be obtained from $\Delta G = \Delta H - T\Delta S = 0$ which after rearrangement leads to

$$T = \frac{\Delta H}{\Delta S} \tag{4.5}$$

The Gibbs function can also be interpreted in the context of efficiency analogue to potential energy in mechanics and the ratio of work done to heat supplied in energy analysis. The Gibbs free energy gives the maximal energy available to perform work, while the enthalpy gives the total amount of energy released due to phase transformations or chemical reactions and the $T\Delta S$ represents the loss term or non-useable energy needed for expansion work. The efficiency is then defined as $\frac{\Delta G}{\Delta H}$.

Similar to the introduction of standard enthalpies of formation (see Sect. 4.2), we can also define *standard Gibbs free energies of formation* (ΔG_f^o) which is the *change in Gibbs free energy for the formation of one mol of that compound from its elements under standard conditions* (STAP, 25 °C and 1 bar). These ΔG_f^o values are tabulated, reported in kJ mole^{-1} and provide information on the stability of compounds. For instance, the ΔG_f^o values of water in liquid and gas forms are -237.1 and -228.6 kJ mol^{-1} respectively, indicating that liquid water is the stable phase at 25 °C and 1 bar pressure.

These ΔG_f^o values can also be used to calculate the *Gibbs free energy of a reaction* $(\Delta_r G^o)$ at 298 K and 1 bar:

$$\Delta_r G^o = \sum v_i \Delta G_f^o(products) - \sum v_i \Delta G_f^o(reactants), \tag{4.6}$$

where v_i is the stoichiometric coefficient of the reaction.

Returning to reaction (Eq. 3.9) discussed before for which we have provided the thermodynamic data (Table 3.1):

$$2H_2(g) + O_2(g) \rightarrow 2H_2O(l)$$

Like the entropy $(\Delta_r S^o)$ and enthalpy $(\Delta_r H^o)$ of reactions, the Gibbs free energy $(\Delta_r G^o)$ can be calculated from the product-minus-reactants (Eq. 4.6).

$\Delta_r G^o = \left[2 \times \Delta G_f^o(H_2O(l))\right] - \left[2 \times \Delta G_f^o(H_2(g)) + 1 \times \Delta G_f^o(O_2(g))\right] = -474.26kJ$,

indicating that water is the stable phase at earth surface conditions.

This Gibbs free energy change $(\Delta_r G^o)$ can also be calculated based on the changes in entropy $(\Delta_r S^o)$ and enthalpy $(\Delta_r H^o)$ of reactions (Eq. 4.3):

$\Delta_r G^o = \Delta_r H^o - T\Delta_r S^o = -571.66 \times 1000 - 298.15 \times -326.68 = -474.26$ kJ

and the results are identical.

The first approach via the products-minus-reactants approach is simpler but is restricted to SATP conditions because ΔG_f^o values are tabulated for SATP conditions, while the second approach via enthalpies and entropies can be used at other temperatures as well. However, calculation of Gibbs free energy of reactions at non-standard temperatures via the second approach only holds if enthalpies and entropies do not vary significantly with temperature. In this introduction course, we make this assumption, being aware that both enthalpy and entropy vary with temperature.

Changes in Gibbs free energy during chemical reactions can be used to perform work, to heat our homes or to support the metabolism of organisms. As an example of the latter, we calculate the $\Delta_r G^o$ for microbially mediated reactions. For aerobic (involving oxygen) and anaerobic (in the absence of oxygen, involving sulphate) degradation of model organic matter CH_2O ($\Delta G_f^o = -128.7$ kJ mol^{-1}) we have the following reactions:

$$CH_2O(s) + O_2(g) \rightarrow CO_2(aq) + H_2O(l)$$

$$2CH_2O(s) + SO_4^{2-}(aq) \rightarrow H_2S(g) + 2HCO_3^-(aq)$$

The ΔG_r^o values for these two reactions are -494 and -205 kJ, respectively, or normalized to one mole of CH_2O: -494 and -102 kJ. First, both aerobic respiration and sulphate reduction provide enough energy needed for growth and maintenance of the microbes involved (about -10 kJ per mol). Second, about five times more energy is liberated during aerobic respiration than during sulphate reduction. Where is the missing energy? It is transferred to the reaction products: the hydrogen sulphide produced can react with oxygen:

$$H_2S(g) + 2O_2(g) \rightarrow H_2SO_4(aq)$$

and the $\Delta_r G^o$ value for this reaction is -329 kJ. Microbial communities are highly efficient in using energy and the product of one organism is often the reactant for another organism, i.e., these reactions are coupled, with the result that most of the energy potentially available is eventually used.

4.1   How Gibbs Free Energy Depends on Conditions

The Gibbs free energy change during a process provides information whether a reaction or phase transformation will be spontaneous at constant temperature and pressure ($\Delta G < 0$), but chemists and earth system scientists are often interested how the natural variables temperature or pressure impact the direction of processes or the stability of certain substances. We will therefore derive an equation, the *fundamental Gibbs equation*, expressing the dependence of Gibbs energy on temperature and pressure. The term fundamental is used because P and T are the basic natural variables. We will use basic calculus for the derivations.

Starting from the definition of the Gibbs free energy (Eq. 4.2),

$$G \equiv H - TS$$

and substituting the formal definition of H (Eq. 2.11):

$$H \equiv U + PV,$$

We arrive at,

$$G = H - TS = U + PV - TS.$$

Next, we take the total differential (i.e., the differential of terms of each variable in turn)

$$dG = dU + PdV + VdP - TdS - SdT \qquad (4.7)$$

Making use of the first law of thermodynamics:

$$dU = q - PdV$$

and the definition of entropy (Eq. 3.5):

$$dS = \frac{q}{T},$$

rewritten as $TdS = q$,
    we arrive at

$$dU = TdS - PdV$$

Substituting this equation into Eq. 4.7 leads to

$$dG = TdS - PdV + PdV + VdP - TdS - SdT$$

Simplifying by cancelling terms, we eventually arrive at the *fundamental Gibbs equation* which expresses the dependence of Gibbs free energy on changes in pressure and temperature:

$$dG = VdP - SdT \qquad (4.8)$$

Another way to quantify this dependency is to use the total differential (see Eq. 1. 9 in Box 1):

$$dG = \left(\frac{\partial G}{\partial P}\right)_T dP + \left(\frac{\partial G}{\partial T}\right)_P dT \qquad (4.9)$$

Comparing these equations, the partial derivative with respect to temperature is minus entropy

$$\left(\frac{\partial G}{\partial T}\right)_P = -S \qquad (4.10)$$

Hence, the entropy determines the temperature sensitivity of the Gibbs free energy. The larger the entropy, the greater the temperature dependence. Moreover, the Gibbs free energy of a compound declines (becomes more negative) with increasing temperature (because S values of compounds are always positive).

Similarly, the partial derivative with respect to pressure is volume

$$\left(\frac{\partial G}{\partial P}\right)_T = V \qquad (4.11)$$

Consequently, the volume of a substance determines how its Gibbs free energy changes with pressure. Solids and liquids have small molar volumes (compared to gases) and the pressure dependence of the Gibbs free energy for solids and liquids is rather small. This is evidently not the case for gases. For an isothermal process, we can rewrite Eq. 4.8 as follows:

$$dG = VdP = \left(\frac{nRT}{P}\right)dP$$

Integrating this equation to find ΔG

$$\Delta G = \int_{P_i}^{P_f} dG = \int_{P_i}^{P_f} \left(\frac{nRT}{P}\right)dP = nRT \int_{P_i}^{P_f} \frac{1}{P}dP = nRT ln\left(\frac{P_f}{P_i}\right) \qquad (4.12a)$$

where we have used the ideal gas law to eliminate V and $\int \frac{dx}{x} = lnx$ for the integration.

This equation expresses the isothermal change in Gibbs free energy of a gas when pressures changes from P_i to P_f.

The change in Gibbs free energy (ΔG) at pressure P can then be expressed relative the standard Gibbs free energy change at 1 bar (ΔG^o)

$$\Delta G = \Delta G^o + nRT ln\left(\frac{P}{1}\right) = \Delta G^o + nRT ln(P) \qquad (4.12b)$$

This latter equation is only valid if pressure is expressed in bar units, but the ratio $\left(\frac{P_f}{P_i}\right)$ is dimensionless and Eq. 4.12a can be used when other units or other reference levels are adopted.

Equation 4.12 has been derived for gases expressed in pressures, but an equivalent expression can be derived when considering ideal liquid or solid solutions. In this case, the concentration [c] in mol L^{-1} ($[c] = \frac{n}{V}$) is the analogue of pressure and the change in Gibbs free energy is:

$$\Delta G = \Delta G^{o} + nRT ln\left[\frac{c}{c^{o}}\right] \tag{13a}$$

where $[c^{o}] = 1$ mol L^{-1} for an ideal solution. For non-ideal solutions, we use *thermodynamic activity* **a** rather than concentration [c]:

$$\Delta G = \Delta G^{o} + nRT ln a \tag{13b}$$

Thermodynamic activities for pure liquids and solids are 1 and the second term then becomes equal to zero and disappears as it should be because $\Delta G = \Delta G^{o}$ for pure phases. Later in the course (Chap. 5) we return to Eqs. 4.12 and 4.13 when we discuss equilibria of solutions.

4.2   Phase Equilibria

In the previous section we have derived the Gibbs free energy equation for a single phase and compound as a function of temperature and pressure. Now we are going to study the relationships between phases and how these depend on pressure and temperature conditions. *A phase is a part of a system which is homogenous and separated from other phases by a definite boundary.* As discussed earlier we deal with gas, liquid, and solid phases. Gases are always present as single phase because they mix well. A liquid containing a single component will only be one phase, but a mixture of liquid may be one phase when well mixed (ethanol in water) or multiphase when they do not mix (water–oil). Solids are often present in multiple phases, even a single component may be present in multiple phases (e.g., carbon in the form of graphite, diamond or fullerene, $CaCO_3$ as calcite or aragonite, Al_2SiO_5 as the minerals kyanite, sillimanite or andalusite).

For phases A and B co-existing in equilibrium we can write:

$$\Delta G = dG_A - dG_B = 0 \tag{4.14}$$

where dG_A and dG_B are the Gibbs free energies of phases A and B. For both phases we can write the Gibbs free energy using the fundamental equation (Eq. 4.8)

$$dG_A = V_A dP - S_A dT \text{ and } dG_B = V_B dP - S_B dT$$

Since phases A and B are in equilibrium, $dG_A = dG_B$, hence

$$V_A dP - S_A dT = V_B dP - S_B dT \tag{4.15}$$

Which after re-arrangement reads

$$V_A dP - V_B dP = S_A dT - S_B dT$$

$$(V_A - V_B)dP = (S_A - S_B)dT$$

Introducing $\Delta V = (V_A - V_B)$ and $\Delta S = (S_A - S_B)$, we can rewrite this as

$$\Delta V dP = \Delta S dT$$

or

$$\frac{dP}{dT} = \frac{\Delta S}{\Delta V} \tag{4.16a}$$

Since we are at equilibrium $\Delta H = T\Delta S$ we arrive at

$$\frac{dP}{dT} = \frac{\Delta H}{T\Delta V} \tag{4.16b}$$

This is the *Clapeyron equation,* and it describes the slope of the line between two phases in equilibrium (also called coexistence curve) in the pressure-temperature domain as a function of changes in molar volume (ΔV) and molar enthalpy (ΔH) or entropy (ΔS). These slopes are usually positive (Fig. 4.1) because entropy and volume changes are positive going from a solid to a liquid or liquid to a gas/vapour.

Fig. 4.1 Generic one component phase diagram. The lines of co-existence between two phases are described by the Clapeyron relation. The triple point is the P–T combination at which solid, liquid and gas co-exist. The critical point is the P–T combination beyond which liquid and vapour have their usual properties

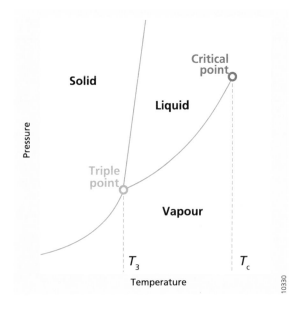

The Clapeyron equation can be used to determine if a phase transition is likely to take place under specific conditions. For instance, considering solids with a typical entropy of fusion (ΔS_{fusion}) of ≈ 22 J mol^{-1} K^{-1} and molar volume change during fusion (ΔV_{fusion})) of $\approx 4 \times 10^{-6}$ m^3 mol^{-1} (i.e., 4 ml per mol), would yield

$$\frac{dP}{dT} = \frac{\Delta S}{\Delta V} = \frac{22}{4 \times 10^{-6}} = 5.5 \times 10^{+6} Pa K^{-1} = 55 bar K^{-1}$$

Accordingly, a pressure increase of ≈ 55 bar is required to change the melting temperature of solids by one degree. The Clapeyron equation is frequently used in solid-earth sciences because it relates the (geo)thermal gradient ($\frac{dT}{dP} = \frac{dT}{dz}$) to changes in volume and enthalpy and provides then an estimate for the temperature T at which minerals are transformed or melting (solidus).

This simple calculation can also be applied to water–ice equilibria although the volume change ΔV going from ice to liquid is negative ($- 1.63$ mL or $- 1.63 \times 10^{-6}$ m^3 for one mol of H_2O), indicating that the P to T slope is negative (Fig. 4.2). Another application is the calculation of freezing point increases with altitude (see Box 3).

While volume changes for solid–liquid transition are rather limited (in the order of mL per mol), this is not the case for gases (about 24 L per mol) and slopes of

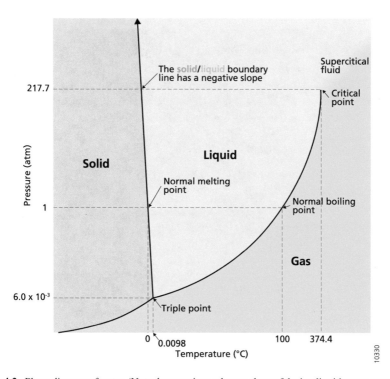

Fig. 4.2 Phase diagram of water. (Note the negative and steep slope of the ice-liquid water transition because of the negative volume change)

liquid–gas co-existence are flatter (Fig. 4.1). The change in volume of the system can thus be approximated with the volume of the gas, i.e. $\Delta V = V_{gas}$. Assuming an ideal gas, we can rewrite Eq. 4.16 to

$$\frac{dP}{dT} = \frac{\Delta H}{T \Delta V} = \frac{\Delta H}{T} \frac{P}{RT} = \frac{\Delta H}{RT^2} P$$

Next, we isolate pressure P on the left-hand side:

$$\frac{dP}{P} = \frac{\Delta H}{RT^2} dT = \frac{\Delta H}{R} \frac{dT}{T^2} \tag{4.17}$$

Making use of $\int \frac{dP}{P} = \ln P$ and integrating from P_1, T_1 to P_2, T_2 we arrive at the *Clausius-Clapeyron* equation:

$$\ln \frac{P_1}{P_2} = -\frac{\Delta H}{R} \left(\frac{1}{T_1} - \frac{1}{T_2} \right) \tag{4.18}$$

The Clausius-Clapeyron equation is very useful for liquid–vapour/gas equilibria such as vapour pressure or boiling points calculation. As an example of the latter, we return to the Himalayas (Box 3).

Box 3 Freezing and Boiling Points in the Himalayas

Consider a mountain high in Himalaya with an atmospheric pressure of 35 kPa (altitude about 7.6 km) and calculate the freezing point increase. The molar enthalpy of melting (ΔH_{fusion}) is 6.01 kJ mol^{-1}, the volume change from ice to liquid is $- 1.63 \times 10^{-6}$ m^3 for one mol of H_2O and the pressure difference (ΔP) is $35 \times 10^3 - 1.013 \times 10^5 = - 6.63 \times 10^4$ Pa.

Using the *Clapeyron equation* (Eq. 4.16b),

$$\frac{dP}{dT} = \frac{\Delta H}{T \Delta V} = \frac{6.01 \times 10^3}{273.15 * -1.63 \times 10^{-6}} = -1.35 \times 10^7 \, Pa\,K^{-1}$$

Hence: $\Delta T = \frac{\Delta P}{-1.35 \times 10^7} \approx 0.005$ K.

The freezing point is thus 0.005 °C higher.

A similar calculation can be made for the lowering of the boiling point using the *Clausius-Clapeyron* equation (Eq. 4.18). The boiling point at 1 atm is 100 °C but it will be lower high in the mountains because of lower atmospheric pressure. Using Eq. 4.18, with $P_1 = 1$ atm $= 1.013 \times 10^5$ Pa and $T_1 = 100$ °C $= 273.15$ K and the standard enthalpy of vaporization of water (ΔH_{vap}) of 40.7 kJ mol^{-1}:

$$\ln\frac{1.013 \times 10^5}{35 \times 10^3} = -\frac{40,700}{8.314}\left(\frac{1}{373.15} - \frac{1}{T_2}\right)$$

we obtain a boiling point of 345 K (= 72 °C) at 7.6 km altitude. It takes much more time to boil an egg high in the mountains because the heat transfer from the water to your food is less.

4.3 Phase Diagrams

Materials exist in the solid phase at high pressure and/or low temperature conditions, but eventually transform into gas/ vapor at low pressures or high temperatures. Phase diagrams are a common way to depict which phase is present at a given P and T (Fig. 4.1). The phases (solid, liquid and gas) are separated by boundary lines, i.e., the lines represent conditions at which two phases co-exist. The Clapeyron relation (Eq. 4.16): $\frac{dP}{dT} = \frac{\Delta H}{T \Delta V}$ gives the slope of these lines.

There are three types of lines: (1) the line between the solid and liquid phase shows how the melting points (temperature of solid to liquid transformation) change with pressure, (2) the line between liquid and vapor phase shows the dependence of boiling points (temperature of liquid to vapour transformation) on pressure, and (3) the line between solid and vapor delineates the transition from a solid to gas (sublimation and condensation). The *triple point* represents the P–T combinations at which the three phases co-exist in equilibrium and *the critical point* represents a combination of temperature and pressure beyond which a gas cannot be liquified regardless of the pressures and temperature. The state of matter beyond the critical point is neither a gas or liquid; it is a supercritical fluid that possesses some typical liquid properties such as high density and ability to act as solvent, but the viscosity and diffusion coefficients are more alike in a gas.

The 'normal' melting point is where a horizontal line at 1 bar or 1 atm crosses the solid–liquid equilibrium lines, and the 'normal' boiling point is crossed at the liquid–gas (vapor) equilibrium lines (Fig. 4.2). These are called normal melting/boiling points because they correspond to Earth-surface conditions, e.g., 0 and 100 °C for the normal melting and boiling points for water. In Box 3/Sect. 4.2 we have calculated how boiling points drop when pressure drops or increases when pressure increases.

The phase boundary lines constrain the freedom of selecting pressure and temperature conditions for a system to be in thermodynamic equilibrium. Within a single phase, one can freely vary P and T independently. At the boundary between two phases, i.e. at the equilibrium lines, either P or T can be chosen, the other one is then fixed. Gibbs has formalized the number of independent variables (= degrees of freedom) to describe a multicomponent system. *Gibbs' phase rule* states that

$$F = C - P + 2 \tag{4.19}$$

where F is degrees of freedom, C is the number of components and P is the number of phases in equilibrium. The + 2 term relates to temperature and pressure. Consider a phase diagram for one single component, e.g. H_2O (C=1) (Fig. 4.2). Within fluid water, or within the domain of ice or water vapor, pressure and temperature can be combined in multiple ways because there are two degrees of freedom (F = 1–1 + 2 = 2). At the interface of water–ice, there is only one degree of freedom (F = 1–2 + 2 = 1), i.e., if temperature is chosen, pressure is fixed or the other way around. The boundary lines between mineral phases are therefore called univariant curves in petrology and mineralogy. At the triple point, ice, water and water vapour co-exist (water is freezing and boiling at the same time) and Gibb's phase rule indicates that there are no degrees of freedom (F = 1–3 + 2 = 0). In other words, the triple point has a fixed pressure and temperature for a component, e.g., for water it is 0.0098 °C and 6×10^{-3} atm.

Phase diagrams are instructive to show that melting (solid to liquid) can occur either through increasing the temperature or through lowering the pressure (except for water because of its negative slope). Similarly, boiling of a liquid can not only occur by increasing the temperature, but also induced by lowering the pressure, i.e., it is possible to boil cold water under low pressure conditions. From the quantitative treatment in Sect. 4.2, it is clear that solid–liquid lines are normally steep because the volume changes involved are small, while liquid–gas equilibrium lines are rather flat because volume changes are large (about three orders of magnitude for ideal gases). Moreover, the liquid–gas equilibrium lines usually show substantial curvature because of the temperature dependence of enthalpies of vaporization. Finally, this course is limited to single component phase diagram, but the same principles, including Gibbs' phase rule, apply to multicomponent systems. However, it can get rather complex.

Part II
Equilibria: Solutions, Minerals, Acid-Base and Redox Reactions

Chapter 5
Introduction to Equilibrium

Abstract This chapter defines equilibrium in mixtures and ideal solutions and introduces the equilibrium constant, its relationship with the Gibbs free energy and its dependence on temperature. Homogenous and heterogenous equilibria are distinguished and solid–gas, liquid–gas and mineral-solution equilibria are presented, including Henry's law and the solubility product.

Keywords Equilibrium constant · Reaction quotient · Solubility product · van 't Hoff equation

Our thermodynamic classes have provided us the tools to determine whether a reaction at Earth surface conditions is spontaneous or not, i.e., whether it has a negative or positive Gibbs free energy. For instance, the dissolution of sodium chloride (kitchen salt) is an endothermic process (consuming heat and cooling the water; $\Delta_r H^o > 0$) but occurs spontaneously because it has a negative Gibbs free energy ($\Delta_r G^o$) of -9 kJ mol^{-1}. Spontaneity of a reaction does not imply that the reaction goes to completion, i.e., equilibrium is not where all reactants are converted into products. While addition of kitchen salt to water initially indeed results in the dissolution of NaCl, no further dissolution is observed after that about 357 gr has been added to 1 kg of water. If more than 357 gr of NaCl were dissolved in water, it would precipitate (a backward reaction). The system water-NaCl has reached *equilibrium*, i.e., the *concentrations of reactants and products remain constant over time*. An equilibrium state is attained as the rate of forward (from reactants to products) and backward (from products to reactants) are equal and there are no further changes in concentrations. In System Earth many reactions can be fully described, or be approximated, by an equilibrium approach: the exchange of gases between water and air, proton transfers (acid-base chemistry) and electron transfers (redox reactions) and the precipitation and dissolution of minerals. For instance, the dissolution of calcium carbonate leads to karst processes and the precipitation of calcium carbonate causes formation of stalactites and stalagmites in caves.

It is therefore useful to deepen our understanding and further develop our quantitative tools for predicting equilibria. In this introduction, we will focus on ideal solution

© The Author(s) 2024

J. J. Middelburg, *Thermodynamics and Equilibria in Earth System Sciences: An Introduction*, SpringerBriefs in Earth System Sciences,
https://doi.org/10.1007/978-3-031-53407-2_5

49

chemistry and equilibria at Earth-surface conditions and leave non-ideal solutions and high-temperature and high-pressure equilibria during formation of magmatic and metamorphic rocks to subsequent courses. Following the introduction of the equilibrium constant and its relation to the Gibbs free energy, we will focus on equilibria in ideal solutions involving minerals, acids and bases, and redox reactions.

5.1   The Equilibrium Constant and Its Relation to the Gibbs Free Energy

Nitrogen dioxide (NO_2) is an atmospheric gas resulting from combustion engines (cars, buses, trucks). It is a deep brown gas that contributes to smog above cities. In the atmosphere, it is in equilibrium with dinitrogen tetroxide (a colorless gas):

$$2NO_2(g) \leftrightarrow N_2O_4(g) \tag{5.1}$$

The transformation of $NO_2(g)$ to $N_2O_4(g)$ causes the color to fade until an equilibrium is established, and this reaction has therefore been well studied in the laboratory. Irrespective of the starting conditions (only $NO_2(g)$, only $N_2O_4(g)$, high $NO_2(g)$ with low N_2O_4 (g), or the other way around), the final composition of the reaction mixture is always the same when expressed as the ratio of product to reactant concentrations (the *law of mass action*). Specifically, in this case the calculated concentration ratio $\frac{N_2O_4(g)}{(NO_2)^2(g)}$ is about 216 at 25 °C and 1 bar, irrespective of starting conditions.

To generalize this phenomenon, we introduce the reaction of α moles A and β moles B to form γ moles C and δ moles D

$$\alpha A + \beta B \rightarrow \gamma C + \delta D \tag{5.2}$$

and define a *reaction quotient* Q:

$$Q = \frac{C^\gamma D^\delta}{A^\alpha B^\beta} \tag{5.3}$$

This reaction quotient is dimensionless and general, i.e., it can always be calculated irrespective whether there is an equilibrium or not. When the system is at equilibrium, the value of Q is called the **equilibrium constant K**:

$$K = \frac{C_{eq}^\gamma D_{eq}^\delta}{A_{eq}^\alpha B_{eq}^\beta} \tag{5.4}$$

The value of K expresses the composition of the system, i.e., concentrations or partial pressures of the reactants and products at equilibrium (taking reaction stoichiometry into account) and has a constant value at a certain temperature and pressure. The equilibrium constant has no unit because they cancel out. Reversing a reaction causes an inversion of K, and multiplying the coefficients (α, β, γ, δ) by a common factor raises the equilibrium constant to the corresponding factor. If the K value is very high, the equilibrium mixture will consist mainly of products C and D. If $K \ll 1$, then reactants will dominate the mixture, while similar quantities of reactant and products are expected if K is around 1, depending on the reaction stoichiometry of course.

$$K \gg 1 \text{ Products dominant}$$
$$K \approx 1 \quad \text{Similar quantities}$$
$$K \ll 1 \text{ Reactants dominant}$$

Comparing the reaction quotient Q and the equilibrium constant K provides information on the direction of change, i.e., whether a process is spontaneous. If $Q > K$, then the net reaction goes from right to left (products to reactant, backward reaction is spontaneous) and if $Q < K$ then the reaction goes from left to right (reactants are transformed into products and the forward reaction is spontaneous). If $Q = K$ then the reaction is in equilibrium.

$$Q > K \text{ backward reaction is spontaneous}$$
$$Q < K \text{ forward reaction is spontaneous}$$
$$Q = K \quad \text{reaction is in equilibrium}$$

In thermodynamics lectures we have seen that spontaneity is related to the Gibbs free energy change at constant pressure and temperature; there should thus be a relation between the reaction quotient Q, the equilibrium constant K, and the sign of the Gibbs free energy change ($\Delta_r G$).

Returning to reaction (5.1), the Gibbs free energy change for the gas reaction ($\Delta_r G$) of $NO_2(g)$ to $N_2O_4(g)$

$$\Delta_r G = \Delta G(N_2O_4(g)) - 2 \times \Delta G(NO_2(g)) \tag{5.5}$$

We use $\Delta_r G$ rather than $\Delta_r G^o$ because the reaction is not occurring under standard conditions (i.e., for a pure gas). During the thermodynamic classes (Eq. 4.12) we have derived that

$$\Delta G = \Delta G^o + RT \ln\left(\frac{P_g}{P_0}\right) \tag{5.6}$$

where R is the universal gas constant, T is absolute temperature, ΔG^o is the standard Gibbs free energy change for a pure gas, P_g is the partial pressure of gas g, and P_0 is the total pressure of 1 bar, respectively. Combining Eqs. (5.5 and 5.6), we then arrive at

$$\Delta_r G = \Delta G^o_f(N_2O_4(g)) + RT \ln\left(\frac{P_{N_2O_4}}{P_0}\right) - 2\left[\Delta G^o_f(NO_2(g)) + RT \ln\left(\frac{P_{NO_2}}{P_0}\right)\right]$$

$$(5.7)$$

Using $\Delta_r G^o = \Delta G^o_f(N_2O_4(g)) - 2 \times \Delta G^o_f(NO_2(g))$, we can rewrite Eq. (5.7) as follows,

$$\Delta_r G = \Delta_r G^o + RT \ln\left(\frac{P_{N_2O_4}}{P_0}\right) - 2RT \ln\left(\frac{P_{NO_2}}{P_0}\right) = \Delta_r G^o + RT \ln\frac{\left(\frac{P_{N_2O_4}}{P_0}\right)}{\left(\frac{P_{NO_2}}{P_0}\right)^2}$$

$$(5.8)$$

Accordingly, the Gibbs free energy of reaction $\Delta_r G$ contains the standard Gibbs free energy change of reaction $(\Delta_r G^o)$ that is independent of the partial pressures of the two gases and a term $RT \ln \frac{\left(\frac{P_{N_2O_4}}{P_0}\right)}{\left(\frac{P_{NO_2}}{P_0}\right)^2}$ that depends on the partial pressure of the reactant and product.

Moreover, this ratio $\frac{\left(\frac{P_{N_2O_4}}{P_0}\right)}{\left(\frac{P_{NO_2}}{P_0}\right)^2}$ is the reaction quotient Q for reaction 5.1, if expressed in partial pressures and a total pressure P_0 of 1 bar.

Generalizing this to any reaction, we obtain

$$\Delta_r G = \Delta_r G^o + RT \ln Q \tag{5.9}$$

stating that the Gibbs free energy change of a reaction is the sum of the standard Gibbs free energy change of the reaction $(\Delta_r G^o)$, calculated from the Gibbs free energy of formations, and the term $RT \ln Q$, where the latter provides a correction term for the actual composition of the mixture.

As an example, consider reaction 5.1 and a reaction vessel at 298 K with 0.350 bar $NO_{2(g)}$ and 0.65 bar $N_2O_{4(g)}$. The standard Gibbs free energy of formations of the two gases are ΔG^0_f 51.3 and 99.8 kJ mol^{-1}, respectively.

$$\Delta_r G = \Delta_r G^o + RT \ln Q$$

$$= 99,800 - 2 \times 51,300 + 8.314 \times 298 \ln \frac{0.65}{0.35^2} = 1330 \text{ J}$$

The $\Delta_r G \neq 0$, the system is thus not at equilibrium. Specifically, the net reaction proceeds towards reactants, i.e., $NO_2(g)$, because $\Delta_r G > 0$. (The same conclusion could be reached using the reaction quotient Q and the equilibrium constant K).

At equilibrium, the reaction quotient is equal to the equilibrium constant ($Q = K$), and the Gibbs free energy of the reaction should be zero ($\Delta_r G = 0$; see thermodynamic lectures, Eq. (4.4)). Consequently,

$$0 = \Delta_r G = \Delta_r G^o + RT \ln K$$

Hence,

$$\Delta_r G^o = -RT \ln K \tag{5.10a}$$

or alternatively written,

$$K = \exp\left(\frac{-\Delta_r G^o}{RT}\right) \tag{5.10b}$$

Equations (5.10a and 5.10b) are completely general and refer to all types of reactions. They provide a direct link between thermodynamic data and theory on the one hand and the composition of equilibrium mixtures on the other. However, Eqs. (5.10a and 5.10b) implies that the mixture composition is expressed in activities rather than concentrations (see Eq. 4.13). In this course, we restrict ourselves to ideal solutions, i.e., ion pairing and interactions between ions and the solvent, are ignored and we express the equilibrium constants and reaction quotients in *concentrations* [c], rather than thermodynamic activities **a**, where $[c] = \frac{P_g}{P_0}$ for a gas ($P_0 = 1$ bar), $[c] = \frac{C_a}{C_0}$ for a solute ($C_0 = 1$ mol L^{-1}) and $[c] = 1$ for solids or the liquid medium. Activities (or effective concentrations) and concentrations are related via activity coefficients. However, the use of concentrations rather than activities implies the need of units for equilibrium constants.

Equilibrium mixture compositions are temperature dependent and there are two ways to correct equilibrium constants for lower or higher temperatures. The first method uses Eq. (5.10a and 5.10b) to link the equilibrium constant K with the $\Delta_r G^o$ at the temperature of interest calculated from the entropy ($\Delta_r S^o$) and enthalpy ($\Delta_r H^o$) of reactions (Eq. 4.3 in thermodynamics). Alternatively, the van 't Hoff equation is used.

Combining, Eq. (5.10a): $\Delta_r G^o = -RT \ln K$.

and Eq. (4.3): $\Delta_r G^o = \Delta_r H^o - T \Delta_r S^o$.

we obtain the *van 't Hoff equation* after rearrangement,

$$\ln K = \frac{\Delta_r S^o}{R} - \frac{\Delta_r H^o}{RT} \tag{5.11}$$

Or in its more useful form going from T_1 to T_2,

$$\ln K_2 = \ln K_1 + \frac{\Delta_r H^o}{R}\left(\frac{1}{T_1} - \frac{1}{T_2}\right) \tag{5.12}$$

where T_1 is usually 298 K (SATP). If the reaction is exothermic ($\Delta_r H^o < 0$), then K decreases with increasing temperature and equilibrium shifts towards more reactants, i.e., to the left. Conversely, equilibria shift to the right and K increases with temperature for an endothermic reaction ($\Delta_r H^o > 0$).

5.2   Heterogenous Equilibria, Including Mineral Solubility

Equilibrium can be established when all reactants and products are in one single phase (*homogenous equilibrium*), or multiple phases are involved (*heterogenous equilibrium*). Acid-base reactions in solution are an example of the former, while mineral dissolution/precipitation reactions involving a solid and a liquid phase are heterogenous. This section will present three types of heterogenous equilibria (solid–gas, liquid–gas and mineral-solution). Homogenous acid-base equilibria will be covered in the next chapter.

Solid–gas equilibria: The production of lime (CaO, s) from limestone ($CaCO_3$, s)

$$CaCO_3(s) \leftrightarrow CaO(s) + CO_2(g)$$

is a heterogenous reaction with an equilibrium constant K defined as

$$K = \frac{[CaO][CO_2]}{[CaCO_3]} = [CO_2]$$

Note that [...] are used to indicate concentrations rather than activities and that solid phase and liquid can be excluded from heterogenous equilibria mass action laws since they have an (activity) value of one. Consequently, the equilibrium concentration of carbon dioxide gas in a system with both solid phase CaO and $CaCO_3$ is independent of the amount of these solid phases, as long as both are present.

Liquid–gas equilibria: The solubility of gases in water is another example of heterogenous equilibrium. The solubility of a gas in water is described using *Henry's law*:

$$C_{gas} = K_H \cdot P_{gas} \tag{5.13}$$

where C_{gas} is the equilibrium concentration of a gas in water (mol kg^{-1}), P_{gas} is the partial pressure (atm) and K_H is the Henry's law constant (mol kg^{-1} atm^{-1}). Henry's law constant is a function of temperature and salinity of the water. Consider the solubility of oxygen in water:

$$O_2(g) + H_2O(l) \leftrightarrow O_2(aq)$$

The corresponding equilibrium constant is $K_{H,O_2} = \frac{O_2(aq)}{P_{O_2}}$.

Figure 5.1 shows the concentrations of dissolved oxygen as a function of temperature for freshwater and ocean water. Oxygen and most other gases dissolve better in freshwater than seawater because of a salting out effect. Dissolved oxygen concentrations in water decline with warming of the water. Cold waters in equilibrium with the atmosphere therefore have higher oxygen contents than warm waters.

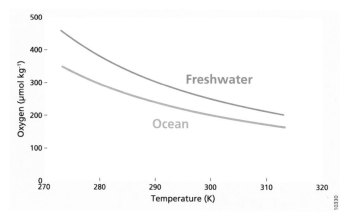

Fig. 5.1 Solubility of oxygen as a function of temperature in freshwater and seawater

Mineral-solution equilibria: The availability of certain minerals constrains the composition of natural waters and vice versa the composition of solutions can determine whether minerals dissolve or precipitate. Mineral equilibria are written as dissolution reactions (i.e., mineral on the left-hand side and ions on the right-hand side) and characterized by a *solubility product*. The solubility product of the mineral fluorite ($CaF_{2(s)}$) is based on the following reaction:

$$CaF_2(s) \leftrightarrow Ca^{2+}(aq) + 2F^-(aq) \tag{5.14}$$

with the following equilibrium constant

$$K = \frac{[Ca^{2+}]_{eq}[F^-]_{eq}^2}{[CaF_2]}$$

where the subscript *eq* indicates that these are equilibrium concentrations. Recalling that solid phases have a (molar fraction) value of one, we can then define the solubility product for fluorite as

$$K_{sp} = [Ca^{2+}]_{eq}[F^-]_{eq}^2 \tag{5.15}$$

The *solubility product* K_{sp} is the equilibrium constant (K) for the special case that a solid phase (mineral) is dissolving. This can be generalized to

$$C_x A_y \leftrightarrow x C_{eq}^{a+} + y A_{eq}^{b-}$$

and

$$K_{sp} = [C^{a+}]_{eq}^x [A^{b-}]_{eq}^y \tag{5.16}$$

Returning to the solubility of fluorite (reaction 5.14), if we know the equilibrium concentrations of dissolved calcium (2×10^{-4} M) and fluoride (1.6×10^{-4} M), we can use Eq. (5.15) to calculate the solubility product as 5.1×10^{-12}. A solubility product is often reported as a pK_{sp} value, i.e. the $-\log_{10}(K_{sp})$, in this case $pK_{sp} = 11.29$.

The same result would be obtained if calculated using Eqs. (5.10a and 5.10b) and thermochemical data with ΔG_f^o values of -1175.6, -553.6, -278.8 kJ mol^{-1} for fluorite, dissolved calcium, and fluoride, respectively. First, one calculates the Gibbs free energy of reaction $(\Delta_r G^o)$:Type equation here.

$$\Delta_r G^o = \left[1 \times \Delta G_f^o \left(Ca^{2+}(aq) \right) + 2 \times \Delta G_f^o \left(F^-(aq) \right) \right]$$
$$- \left[\Delta G_f^o(CaF_2(s)) \right] = 64.4 \text{ kJ}$$

and then uses Eqs. (5.10a and 5.10b):

$$K = \exp\left(\frac{-\Delta_r G^o}{RT} \right) = \exp\left(\frac{-64400}{8.31 \cdot 298} \right) = 5 \times 10^{-12} \rightarrow pKsp = 11.29$$

Accordingly, solubility products link thermochemistry with solution chemistry for solutions in equilibrium.

Solubility products can also be used to constrain the composition of natural waters. Consider a pool in equilibrium with the minerals fluorite (CaF$_2$, $pK_{sp} = 11.29$), calcite (CaCO$_3$, $pK_{sp} = 8.48$) and dolomite (CaMg(CO$_3$)$_2$, $pK_{sp} = 19$), knowing that the dissolved fluoride concentration is 10^{-4} M. From the fluorite equilibrium, we then obtain a dissolved calcium concentration of 5.13×10^{-4} M, next the calcite equilibrium can be used to derive the carbonate ion concentration 6.46×10^{-6} M and finally the dissolved magnesium concentration is calculated making use of the solubility product of dolomite (4.68×10^{-6} M). This example clearly shows that sharing common ions has major implications for the solubility of minerals.

The precipitation and dissolution of minerals depends on the *saturation state* of that mineral. For the dissolution of calcium carbonate in the form of calcite:

$$CaCO_3(s) \leftrightarrow Ca^{2+}(aq) + CO_3^{2-}(aq) \tag{5.17}$$

we first define the *ion product* (*IP*), which is a special form of the *reaction quotient* Q for mineral equilibria.

$$IP = \left[Ca^{2+} \right]\left[CO_3^{2-} \right] \tag{5.18}$$

with, in this case, a unit of mol^2 kg^{-2}, and then the degree of saturation or saturation state (Ω)

$$\Omega = \frac{\left[Ca^{2+} \right]\left[CO_3^{2-} \right]}{\left[Ca^{2+} \right]_{eq}\left[CO_3^{2-} \right]_{eq}} = \frac{\left[Ca^{2+} \right]\left[CO_3^{2-} \right]}{K_{sp}} = \frac{IP}{K_{sp}} \tag{5.19}$$

Calcite will dissolve if $\Omega < 1$, i.e., in undersaturated water, and precipitate from solution if $\Omega > 1$, i.e., if waters are supersaturated with respect to the mineral.

Chapter 6
Acid-Base Equilibria

Abstract This chapter presents acid-base equilibria and simple methods to calculate the pH of solutions. The inorganic carbon system in water is introduced, including the concept of alkalinity, and methods to solve carbon dioxide equilibria in water are discussed. Earth system science relevant examples such as rainwater, surface waters in equilibrium with calcium carbonate minerals, soda lakes and the physical chemistry of karst systems are presented.

Keywords Acid · Base · Dissolved inorganic carbon · Alkalinity · Karst · Soda lake · Rainwater · Ocean acidification

6.1 Introduction

Although protons are produced and consumed in many chemical processes, most natural waters fall within a narrow range of pH values of 6–9 because of interactions with minerals and transfer of protons (i.e., hydrogen ions) among acids and bases in solution. Most proton transfer reactions in solution are very fast (i.e., reach an equilibrium in milliseconds) and a (homogenous) equilibrium description is thus appropriate. A proton does not exist as such in water because it reacts fast and strongly with water to form a hydrated proton (hydronium), H_3O^+, which in turn associates with hydrogen bonds to additional water molecules. It is macroscopically not possible to distinguish between the various hydrated proton species and we will therefore use H^+ and H_3O^+ interchangeably and refer to H^+ as the proton or hydrogen ion in solution.

An **acid** is a substance that dissociates in water releasing protons into solution, i.e., it is a *proton donor*, while a **base** *accepts protons*. Chemical species whose formulas differ only by one hydrogen ion are called conjugate acid-base pairs. An acid HA dissolved in water dissociates into a proton and a base A^-:

$$HA(aq) + H_2O(l) \leftrightarrow H_3O^+(aq) + A^-(aq) \tag{6.1}$$

© The Author(s) 2024

J. J. Middelburg, *Thermodynamics and Equilibria in Earth System Sciences: An Introduction*, SpringerBriefs in Earth System Sciences, https://doi.org/10.1007/978-3-031-53407-2_6

for which we can write the general equilibrium constant K

$$K = \frac{[H_3O^+][A^-]}{[HA][H_2O]}$$

Noting that the liquid medium has a (activity) value of 1, we can then define the *acid dissociation or acidity constant*

$$K_a = \frac{[H^+][A^-]}{[HA]} \tag{6.2}$$

A similar treatment can be followed for base B added to water

$$B(aq) + H_2O(l) \leftrightarrow OH^-(aq) + BH^+(aq) \tag{6.3}$$

with the *base dissociation or basicity constant*

$$K_b = \frac{[OH^-][BH^+]}{[B]}, \tag{6.4}$$

but we will elaborate it from the acid side of the story.

The extent of dissociation, i.e., the proton release by the acid HA and its transfer to water acting as the base in Reaction 6.1, depends on the strength of the acid, i.e., the value of K_a. Strong acids nearly fully dissociate into H^+ and A^- and almost no HA remains in solution. Weak acids dissociated partially and after equilibration HA, H^+ and A^- are found in solution. Acids stronger than H_3O^+ are called strong acids and their conjugated bases are very weak. Weak acids are less strong than H_3O^+ but stronger than H_2O (Table 6.1) and their conjugated bases are also weak.

Reactions 6.1 and 6.3 show that water can act as a proton donor as well as proton acceptor. The dissociation or *self-ionization of water*,

$$2H_2O(l) \leftrightarrow H_3O^+(aq) + OH^-(aq) \tag{6.5}$$

leads to the *ion-product constant* for water, i.e., the equilibrium constant for self-ionisation of water,

$$K_w = [H_3O^+][OH^-] = 10^{-14} \tag{6.6}$$

Some acids contain multiple protons, and these are called polyprotic acids (e.g. carbonic acid, phosphoric acid). These polyprotic acids dissociate stepwise, and each dissociation step is characterized by its own acid dissociation constant. In solutions containing multiple acids and bases, protons are transferred from the stronger acid to the stronger base to yield the weaker acid and base.

For notational simplicity, we will drop the phase descriptions *aq* for solutes and *l* for liquid water, but the *s* for solids is retained. Acid and base dissociation constants

Table 6.1 Acid and base dissociation constants of conjugated acid-base pairs

Name	Acid	pKa	Base	pKb
Hydrochloric acid	HCl	Strong	Cl^-	Very weak
Sulfuric acid	H_2SO_4	Strong	HSO_4^-	Very weak
Nitric acid	HNO_3	Strong	NO_3^-	Very weak
Proton	H_3O^+		H_2O	
Chromic acid	H_2CrO_4	0.74	$HCrO_4^-$	13.26
Iodic acid	HIO_3	0.78	IO_3^-	13.22
Sulfurous acid	H_2SO_3	1.85	HSO_3^-	12.15
Bisulfate	HSO_4^-	1.99	SO_4^{2-}	12.01
Phosphoric acid	H_3PO_4	2.16	$H_2PO_4^{2-}$	11.84
Hydrofluoric acid	HF	3.2	F^-	10.8
Nitrous acid	HNO_2	3.25	NO_2^-	10.75
Hydrogen selenide	H_2Se	3.89	HSe^-	10.11
Acetic acid	CH_3CO_2H	4.76	$CH_3CO_2^-$	9.24
Carbonic acid	H_2CO_3	6.35	HCO_3^-	7.65
Bichromate	$HCrO_4^-$	6.49	CrO_4^{2-}	7.51
Hydrogen sulfide	H_2S	7.05	HS^-	6.95
Bisulfite	HSO_3^-	7.2	SO_3^{2-}	6.8
Dihydrogenphosphate	$H_2PO_4^{2-}$	7.21	HPO_4^{2-}	6.79
Boric acid	H_3BO_3	9.27	$H_2BO_3^-$	4.73
Ammonium	NH_4^+	9.3	NH_3	4.7
Bicarbonate	HCO_3^-	10.33	CO_3^{2-}	3.67
Biselenide	HSe^-	11	Se^{2-}	3
Biphosphate	HPO_4^{2-}	12.32	PO_4^{3-}	1.68
Water	H_2O		OH^-	
Ammonia	NH_3	Very weak	NH_2^-	Strong
Hydroxy ion	OH^-	Very weak	O^{2-}	Strong

are often presented as pK values ($-\log K$), and $[H^+]$ and $[OH^-]$ concentrations are traditionally presented as pH and pOH values ($-\log [H^+]$ and $-\log [OH^-]$). The pH scale is logarithmic implying that a 0.3 increase corresponds to a halving of concentrations ($-\log(\frac{1}{2}) = 0.3$) and that the absolute change depends on the value: from pH 5 to 5.3 corresponds to a 5 µM proton concentration decrease from 10 to 5 µM, while a pH change from 6 to 6.3 relates to 0.5 µM change in $[H^+]$ from 1 to 0.5 µM.

pH calculations: When calculating the pH of a solution in which an acid is dissolving, the first step is to evaluate whether the acid is strong ($pK_a < 0$) or weak.

A strong acid will fully dissociate, i.e., all the acid HA added to the solution will be converted into H^+ and A^- and the eventual pH $= -\log$ [HA]. A 0.1 M solution with the strong acid HCl will thus have a pH of 1.

The pH of a solution of a weak acid HA (with concentration C_a) dissolving in pure water can be calculated with a (R)ICE table.

(**Reaction**)	[HA]	$\left[H^+\right]$	$\left[A^-\right]$
Initial	C_a	≈ 0	0
Change	$-x$	x	x
Equilibrium	$C_a - x$	x	x

The first row is for the reaction: HA $\leftrightarrow H^+ + A^-$

The second row specifies the initial conditions: the concentration of the acid added (C_a) that is initially undissociated so that $\left[A^-\right]$ is zero and $\left[H^+\right]$ is close to zero (10^{-7} for pure water).

The third row gives the change due to reaction; x amount of [HA] is consumed and $x\left[H^+\right]$ and $x\left[A^-\right]$ are produced.

The fourth row, equilibration, is simply the sum of rows two and three.

Next, the acid dissociation constant (K_a) and the concentrations in the equilibrium line in the (R)ICE table are combined in Eq. 6.2.

$$K_a = \frac{\left[H^+\right]\left[A^-\right]}{[HA]} = \frac{x \cdot x}{C_a - x}$$

which can be rewritten as a quadratic equation:

$$x^2 + K_a x - K_a C_a = 0$$

which can be solved for x to obtain the proton concentration, and thus the pH $= -\log(x)$. If $C_a > 1000$ times K_a, then x is very small relative to C_a, and $C_a - x \approx C_a$. The proton concentration can then be calculated from $\left[H^+\right] = \sqrt{K_a C_a}$.

As an example, calculate the pH of 0.1 M Acetic acid solution with a pK_a value of 4.76. The corresponding (R)ICE table reads

(**Reaction**)	[HAc]	$\left[H^+\right]$	$\left[Ac^-\right]$
Initial	0.1	≈ 0	0
Change	$-x$	X	x
Equilibrium	$0.1 - x$	X	x

and the resulting pH $= 2.88$.

Common ions, i.e., substances that are both in the acid/base system as well as in the solution, shift the equilibrium and thus the pH of the solution. Consider the above example, but now with 0.05 M Na-acetate as well in the solution. Na-acetate has a very high dissociation constant and is fully dissociated, i.e., it is present in the form of acetate $[Ac^-]$ and sodium ion $[Na^+]$. Sodium ions are not involved in proton transfer reaction and can be ignored. The (R)ICE table must be modified for the initial conditions to accommodate for the presence of acetate ions.

(Reaction)	[HAc]	$[H^+]$	$[Ac^-]$
Initial	0.1	≈ 0	0.05
Change	$-x$	x	x
Equilibrium	$0.1 - x$	x	$0.05 + x$

The corresponding equilibrium

$$K_a = \frac{[H^+][Ac^-]}{[HAc]} = \frac{x \cdot (x + 0.05)}{0.1 - x}$$

and quadratic equation

$$x^2 + (0.05 + K_a)x - 0.1\,K_a = 0$$

results in a pH of 4.46. The addition of the common ion acetate shifts the equilibrium to the left and the proton concentration declines (and pH increases).

Buffer solutions: Solutions containing weak acids with their conjugated bases are buffer solutions because they resist drastic changes in pH. Quantifying the buffering capacity of natural waters containing multiple acid-base systems can be quite cumbersome. Therefore, we restrict ourselves to the *Henderson-Hasselbalch* equation for a single acid-base system to illustrate the principle.

Starting with re-arranging Eq. 6.2

$$K_a = \frac{[H^+][A^-]}{[HA]}$$

to isolate the proton concentration on the left-hand side

$$\frac{1}{[H^+]} = \frac{[A^-]}{K_a[HA]} = \frac{1}{K_a}\frac{[A^-]}{[HA]}$$

followed by taking the logarithms

$$pH = -\log K_a + \log \frac{[A^-]}{[HA]} = pK_a + \log \frac{[A^-]}{[HA]} \tag{6.7}$$

we have derived the *Henderson-Hasselbalch equation*.

Dissolution of carbon dioxide in water results in the formation of carbonic acid and this provides most of the buffering of natural waters. In the next section we will elaborate the carbon dioxide-water system in quite some detail, here we illustrate the buffering provided by the carbonic acid-bicarbonate system:

$$H_2CO_3 \leftrightarrow H^+ + HCO_3^-$$

Consider a system with initial H_2CO_3 and HCO_3^- concentrations of 1 mM each and a pH of 6.35, i.e., the pK_a value. Using the Henderson-Hasselbalch equation, we obtain

$$pH = pK_a + \log\frac{\left[HCO_3^-\right]}{[H_2CO_3]} = 6.35 + \log\frac{\left[10^{-3}\right]}{\left[10^{-3}\right]} = 6.35$$

Addition of $10~\mu$ mol strong acid to 1 liter would lower HCO_3^- and increase H_2CO_3 concentrations, but hardly impact the pH

$$pH = 6.35 + \log\frac{\left[10^{-3} - 10^{-5}\right]}{\left[10^{-3} + 10^{-5}\right]} = 6.34$$

The ΔpH is 0.01 at the initial pH of 6.35, but it would be ≈ 0.03 at a pH of 5.35 and ≈ 0.31 at a pH of 4.35. Accordingly, a buffer is optimal close to its pK_a value, but still limits proton concentration changes to within a factor of two at pH values two units higher or lower than the pK_a. Carbonic acid is a diprotic acid, i.e. a polyprotic acid with two protons, with pK_a values of 6.35 and 10.33. It thus has the potential to buffer natural waters over a wide range of pH values.

6.2   Carbon Dioxide Equilibria

Carbon dioxide is a gas that exchanges between water and air. When dissolved in water, it reacts with water to form carbonic acid, which in turn dissociates into bicarbonate and carbonate ions. This carbon-dioxide-water system provides much of the buffering capacity for natural waters, governs the pH of most natural waters, is pivotal to precipitation and dissolution of carbonate minerals, and governs the uptake of (anthropogenic) carbon by the ocean.

The carbon-dioxide-water system is described by a few equilibria. Carbon dioxide gas dissolves in water and will then react with water to form carbonic acid (H_2CO_3) according to

$$CO_2(aq) + H_2O \leftrightarrow H_2CO_3(aq) \tag{6.8}$$

This equilibrium lies rather far to the left, hence most of the dissolved carbon dioxide remains in the form of $CO_2(aq)$, and we can analytically not distinguish between $H_2CO_3(aq)$ and $CO_2(aq)$; these two pools are therefore lumped together and termed carbonic acid (H_2CO_3) in this course. Other treatments of carbon-dioxide-water system sometimes use the terms $H_2CO_3^*$ or CO_2^* for the lumped pool. Accordingly, we can re-write Eq. 6.8 to directly link atmospheric carbon dioxide and carbonic acid as follows:

$$CO_{2(g)} + H_2O \leftrightarrow H_2CO_3 \tag{6.9}$$

with the equilibrium expression based on Henry's law (Eq. 5.13):

$$[H_2CO_3] = K_H \cdot P_{CO_2} \tag{6.10}$$

where P_{CO_2} is the partial pressure of carbon dioxide in equilibrium with water and K_H is Henry's Law constant (mol kg^{-1} atm^{-1}) with a pK_H value of 1.47 at 25 °C in pure water. Carbonic acid is a weak acid that partly dissociates into a proton and bicarbonate (HCO_3^-), the latter being another weak acid that dissociates into another proton and carbonate (CO_3^{2-}). The relevant reactions are

$$H_2CO_3 \leftrightarrow H^+ + HCO_3^- \tag{6.11}$$

and

$$HCO_3^- \leftrightarrow H^+ + CO_3^{2-} \tag{6.12}$$

with the corresponding equilibria and equilibrium constants (in freshwater and at 25 °C)

$$K_1 = \frac{[H^+][HCO_3^-]}{[H_2CO_3]} = 10^{-6.35} \tag{6.13}$$

$$K_2 = \frac{[H^+][CO_3^{2-}]}{[HCO_3^-]} = 10^{-10.33} \tag{6.14}$$

Accordingly, the carbon-dioxide-water system is characterized by five aqueous species: $[H^+]$, $[OH^-]$, $[H_2CO_3]$, $[HCO_3^-]$ and $[CO_3^{2-}]$ that are linked via three equilibrium relations (the self-ionization of water, Eq. 6.6, and the first and second acid-base equilibria, Eqs. 6.13 and 6.14) with known constants. To calculate the abundance of the five species, we must solve the system and for this we need more information, specifically, two more relations.

Various types of information can be used. These can either be information on the measured or known concentration of the aqueous species (e.g., known pH), or additional relations. For a system open to exchange with the air, Henrys' law (Eq. 6.10) provides a link of carbonic acid (H_2CO_3) to the partial pressure of carbon dioxide in the air (P_{CO_2}). Solubility equilibria involving carbonate minerals can constrain the carbonate ion concentration. Natural waters are uncharged, and we can thus use the charge balance equation as an additional constraint: the positive charge of protons should be balanced by the negative charge of hydroxide, bicarbonate and carbonate ions.

$$[H^+] = [OH^-] + [HCO_3^-] + 2[CO_3^{2-}] \qquad (6.15)$$

Note that the carbonate ion is counted twice in a charge balance because of its double charge. Alternatively, one can define a proton balance equation, a mass balance for protons

$$[H^+] = [H^+]_{H_2O} + [H^+]_{H_2CO_3} \qquad (6.16a)$$

or its equivalent

$$[H^+] = [OH^-] + [HCO_3^-] + 2[CO_3^{2-}], \qquad (6.16b)$$

This proton conservation equation balances excess protons on the left-hand side with the recipe on the right-hand side and is in this case identical to the charge balance (Eq. 6.15). However, the proton conservation and charge balance equations can be different in more complex solutions such as seawater.

Unfortunately, not all species of the system can be measured directly, but there is also no need because they are interlinked via equilibrium, mass balance and charge conservation equations. The four parameters most often measured are: the pH, the partial pressure of (P_{CO_2}), the dissolved inorganic carbon content and the titration alkalinity. The *total dissolved inorganic carbon content*, abbreviated as C_T or DIC (dissolved inorganic carbon) is defined as:

$$DIC = C_T = [H_2CO_3] + [HCO_3^-] + [CO_3^{2-}]. \qquad (6.17)$$

The titration alkalinity is the excess of proton acceptors over donors of a solution and is normally derived from an acidimetric titration. *Alkalinity* (TA) for our simple system is defined as:

$$TA = [OH^-] + [HCO_3^-] + 2[CO_3^{2-}] - [H^+] \qquad (6.18)$$

This alkalinity definition is rooted in the charge and proton balances presented earlier (Eqs. 6.15 and 6.16). Carbonate alkalinity ($CA = [HCO_3^-] + 2[CO_3^{2-}]$) predominates alkalinity of natural waters because of the presence of carbon dioxide in the atmosphere.

6.3 Solving Carbon Dioxide Equilibria

The previous section has provided the tools to solve carbon dioxide equilibria. However, solving carbon dioxide equilibria in natural waters, particularly in seawater, can be challenging because of the presence of multiple acid-base systems and other species interacting with the carbonate system. Fortunately, there are multiple computer programs available to assist us. Below we present a few simple, Earth-sciences relevant, cases that are instructive and can be done without such computer programs. First, it is important to identify whether the system is *open* (to an atmosphere with known composition) or *closed* (when interactions with the atmosphere are negligible). The open system approach applies to surface waters of streams, rivers, lakes, reservoirs, estuaries, and the surface ocean, while a closed approach is more appropriate for groundwater and subsurface waters in lakes and in the ocean interior that are isolated from the atmosphere. Second, calculations for problems with *known pH* are usually simpler than those without a priori knowledge on pH, and we therefore start with the former.

Case 1 The pH is known

This type of problems is most conveniently solved using the *ionization fraction* approach. The derivation of the ionization fraction starts with rewriting the first carbonic acid equilibrium (Eq. 6.13) to isolate bicarbonate on the left-hand side:

$$K_1 = \frac{[H^+][HCO_3^-]}{[H_2CO_3]} \Rightarrow [HCO_3^-] = \frac{K_1[H_2CO_3]}{[H^+]} \tag{6.19}$$

Next, we substitute this equation in the second carbonic acid equilibrium (Eq. 6.14)

$$K_2 = \frac{[H^+][CO_3^{2-}]}{[HCO_3^-]} \Rightarrow [CO_3^{2-}] = \frac{K_2[HCO_3^-]}{[H^+]} \tag{6.20}$$

and we obtain after re-arrangement an expression with the carbonate ion isolated on the left-hand side:

$$[CO_3^{2-}] = \frac{K_1K_2[H_2CO_3]}{[H^+]^2} \tag{6.21}$$

Combining these equations with the definition of the dissolved inorganic carbon content (Eq. 6.17)

$$\begin{aligned} DIC &= [H_2CO_3] + [HCO_3^-] + [CO_3^{2-}] \\ &= [H_2CO_3] + \frac{K_1[H_2CO_3]}{[H^+]} + \frac{K_1K_2[H_2CO_3]}{[H^+]^2} \\ &= [H_2CO_3]\left(1 + \frac{K_1}{[H^+]} + \frac{K_1K_2}{[H^+]^2}\right) \end{aligned}$$

Introducing the *ionization fraction* and α

$$\alpha = \left(1 + \frac{K_1}{[H^+]} + \frac{K_1 K_2}{[H^+]^2}\right) \tag{6.22}$$

we can express carbonic acid, bicarbonate and carbonate concentrations as a function of DIC, K_1, K_2 and $[H^+]$:

$$[H_2CO_3] = \frac{DIC}{\alpha} \tag{6.23a}$$

$$[HCO_3^-] = \frac{DIC \cdot K_1}{\alpha \cdot [H^+]} \tag{6.23b}$$

$$[CO_3^{2-}] = \frac{DIC \cdot K_1 \cdot K_2}{\alpha \cdot [H^+]^2} \tag{6.23c}$$

This ionization approach can not only be used for closed systems with known DIC, but also for open systems (without a priori information on DIC, but with known pH). In an open system, the carbonic acid concentration is set by Henry's Law (Eq. 6.10), and via Eq. 6.23a one can then estimate DIC first and subsequently the bicarbonate and carbonate ion concentrations.

Figure 6.1 shows the distribution of species contributing to DIC as a function of pH, a Bjerrum plot, calculated using the ionization fraction approach. Carbonic acid is the dominant species at pH values below the pK_1, bicarbonate dominates between the pK_1 and pK_2 values and the carbonate ion dominates at pH values above the pK_2. Moreover, the pH dependent distributions of carbonic acid, bicarbonate and carbonate differ between freshwater and seawater because the many ion interactions, i.e., non-ideal behavior, in seawater shift equilibria to lower pH values.

Case 2 Rainwater pH: an open system

Carbon dioxide in rainwater is in equilibrium with the atmosphere, and we have five unknown aqueous species: $[H^+]$, $[OH^-]$, $[H_2CO_3]$, $[HCO_3^-]$ and $[CO_3^{2-}]$. We can start the calculation using Eq. 6.10 from the known atmospheric partial pressure of carbon dioxide (420 µatm) and Henry's Law constant (3.4×10^{-2} mol kg^{-1} atm^{-1}) to arrive at a carbonic acid concentration of 1.42×10^{-5} M ($= 10^{-4.85}$). Next, we rearrange the first equilibrium relation:

$$K_1 = \frac{[H^+][HCO_3^-]}{[H_2CO_3]} \Rightarrow [HCO_3^-] \cdot [H^+] = K_1 \cdot [H_2CO_3]$$
$$= 10^{-6.35} \times 10^{-4.85} = 10^{-11.20}$$

and see that the product of $[HCO_3^-] \cdot [H^+] = 10^{-11.20}$. Next, we consider the charge balance (Eq. 6.15),

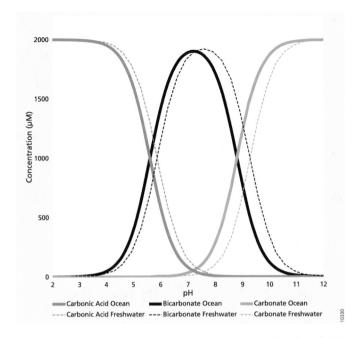

Fig. 6.1 Distribution of carbonic acid, bicarbonate and carbonate as a function of pH in ocean and freshwater for a total DIC of 2000 μM and a temperature of 15 °C

$$\left[H^+\right] = \left[OH^-\right] + \left[HCO_3^-\right] + 2\left[CO_3^{2-}\right]$$

and realize that $\left[HCO_3^-\right] \gg \left[CO_3^{2-}\right]$ and $\left[H^+\right] \gg \left[OH^-\right]$, because after the addition of carbon dioxide pH should be lower than 7. The charge balance can thus be simplified to

$$\left[H^+\right] \approx \left[HCO_3^-\right] \approx 10^{-5.6} \Rightarrow pH = 5.6$$

Pure rainwater (not impacted by atmospheric dust or pollution) is thus slightly acidic. Increasing atmospheric carbon dioxide levels to 1000 or 5000 μatm would lower rainwater pH values to 5.4 and 5.06, respectively.

Humans have released large quantities of SO_2 to the atmosphere due to burning of sulfur bearing fossil fuels for industrial activities and transport. In the air some of the SO_2 gas is oxidized to SO_3 and both these gases have dissolved in rain. This had led to the production of sulfurous (H_2SO_3) and sulfuric acid (H_2SO_4). As a result, pH in rain declined to values down to 3, i.e., *acid rain*, in some industrial areas. Such low pH values have consequences for mineral equilibria, availability of essential nutrients, and organisms' physiology. In the 1980/90s environmental measures such as sulfur removal from exhausts have been implemented resulting in a gradual return to more natural pH values, but some of the impacted ecosystems (e.g., forests) are still recovering from that perturbation.

Case 3 The pH of a solution in equilibrium with calcium carbonate and the atmosphere

This case would correspond to a pool or shallow lake in a carbonate rock setting, or a coastal ocean with carbonate minerals at the seafloor. It also applies to travertine formation in rivers or stalagmite and stalactite formation in caves (see Box 4). The system is characterized by six unknown aqueous species: $[H^+]$, $[OH^-]$, $[H_2CO_3]$, $[HCO_3^-]$ and $[CO_3^{2-}]$ as in the previous case of rainwater, plus $[Ca^{2+}]$ from the equilibrium with carbonate minerals. Consequently, we need six equations to solve the problem. The first four are the self-ionization of water (Eq. 6.6), the first and second acid-base equilibria (Eqs. 6.13 and 6.14) and Henry's Law combined with the P_{CO_2} (Eq. 6.10). Since the water bodies are in contact and equilibrium with calcium carbonate minerals, the solubility product (Eq. 5.16) provides the fifth constraint:

$$K_{sp} = [Ca^{2+}][CO_3^{2-}] \qquad (6.24)$$

The charge balance provides the sixth, final constraint.

$$[H^+] + 2[Ca^{2+}] = [OH^-] + [HCO_3^-] + 2[CO_3^{2-}] \qquad (6.25)$$

Notice that $[Ca^{2+}]$ and $[CO_3^{2-}]$ are counted twice because of their double charge.

With six unknown and six equations, the problem is well posed, but the mathematical solution is somewhat cumbersome. We will first present the full solution and then show that by making use of chemical insight, we can simplify the solution.

Full solution: Since it is an open system in equilibrium with the atmosphere, we can start with Eq. 6.10:

$$[H_2CO_3] = K_H \cdot P_{CO_2}$$

and insert it in Eqs. 6.19 and 6.21 to obtain expressions for bicarbonate and carbonate ions:

$$[HCO_3^-] = \frac{K_1[H_2CO_3]}{[H^+]} = \frac{K_1 K_H \cdot P_{CO_2}}{[H^+]} \qquad (6.26)$$

$$[CO_3^{2-}] = \frac{K_1 K_2[H_2CO_3]}{[H^+]^2} = \frac{K_1 K_2 K_H \cdot P_{CO_2}}{[H^+]^2} \qquad (6.27)$$

Using Eqs. 6.24, 6.6, 6.26, and 6.27, we can rewrite the charge balance Eq. (6.25):

$$[H^+] + 2\frac{K_{sp}}{\frac{K_1 K_2 K_H \cdot P_{CO_2}}{[H^+]^2}} = \frac{K_w}{[H^+]} + \frac{K_1 K_H \cdot P_{CO_2}}{[H^+]} + 2\frac{K_1 K_2 K_H \cdot P_{CO_2}}{[H^+]^2} \qquad (6.28)$$

Multiplying by $[H^+]^2$ and re-arranging so that all terms are on the left-hand side results in

$$\left[H^+\right]^3 + 2\frac{K_{sp}\left[H^+\right]^4}{K_1 K_2 K_H \cdot P_{CO_2}} - K_w\left[H^+\right] - K_1 K_H \cdot P_{CO_2}\left[H^+\right]$$
$$- 2K_1 K_2 K_H \cdot P_{CO_2} = 0 \tag{6.29}$$

This is a fourth-order polynomial in $\left[H^+\right]$ that can be solved by trial-and-error or numerical techniques.

Approximate solution: For a system open to the atmosphere, the final pH is likely close to neutral (7 ± 2), and we can thus simplify the charge balance (6.25) to:

$$2\left[Ca^{2+}\right] = \left[HCO_3^-\right] \tag{6.30}$$

because protons and hydroxide ions are on the order of 10^{-5} to 10^{-9} M and $\left[HCO_3^-\right] > \left[CO_3^{2-}\right]$(see Fig. 6.1). Next, we rewrite the second acid equilibrium (Eq. 6.14) to isolate carbonate on the left-hand side:

$$K_2 = \frac{\left[H^+\right]\left[CO_3^{2-}\right]}{\left[HCO_3^-\right]} \Rightarrow \left[CO_3^{2-}\right] = \frac{K_2\left[HCO_3^-\right]}{\left[H^+\right]} \tag{6.31}$$

Next, we substitute Eqs. 6.30 and 6.31 into the solubility Eq. 6.25

$$K_{sp} = \left[Ca^{2+}\right]\left[CO_3^{2-}\right] \Rightarrow K_{sp} = 0.5\left[HCO_3^-\right]\frac{K_2\left[HCO_3^-\right]}{\left[H^+\right]} = 0.5\left[HCO_3^-\right]^2 \frac{K_2}{\left[H^+\right]} \tag{6.32}$$

As above, equation (6.26) is used to express bicarbonate in terms of P_{CO_2} and $\left[H^+\right]$, and (6.32) is then rewritten as

$$K_{sp} = 0.5\left[\frac{K_1 K_H P_{CO_2}}{\left[H^+\right]}\right]^2 \frac{K_2}{\left[H^+\right]} \tag{6.33}$$

Or after some further re-arrangements to isolate the proton concentration on the left-hand side:

$$\left[H^+\right]^3 = P_{CO_2}{}^2 \left[\frac{K_1^2 K_2 K_H^2}{2K_{sp}}\right] \tag{6.34}$$

Solving this equation for a P_{CO_2} of 420 μatm and using the standard values for pK_1 (6.35), pK_2 (10.33), pK_H (1.47) and the solubility product of calcite ($pK = 8.48$), yields a pH value of 8.18, consistent with observations for freshwaters in carbonate terrains. Moreover, it is also very similar to that of modern seawater despite seawater being a non-ideal solution.

A more accurate way to calculate the pH for seawater in equilibrium with calcite and the modern atmosphere is to use equilibrium constants specific for seawater at

25 °C: $pK_1^{sw}(5.87)$, pK_2^{sw} (8.76), pK_H^{sw} (1.54) and the solubility product of calcite ($p\ K_{sp}^{sw} = 6.36$). These seawater specific constants are shifted towards lower pK values, in other words higher K values (see Fig. 6.1). Modern seawater is estimated to have a pH of 8.09, consistent with observations.

Atmospheric carbon dioxide levels have varied over the Earth's history and Table 6.2 shows the pH value expected for freshwater and marine waters in equilibrium with calcite and the atmosphere at 25 °C for glacial, pre-industrial, modern, Eocene, and projected levels in 2100 for two global warming scenarios (2 °C and 4 °C).

Table 6.2 clearly shows that pH declines with increasing P_{CO_2} levels. This *ocean acidification* is also known as the *other* CO_2 *problem*. The above calculations are based on full equilibration of seawater with both the atmosphere and calcium carbonate in the form of calcite. However, in the ocean, carbonate minerals are at the bottom of the ocean (depth of km's) while gas exchange, and thus CO_2 invasion, occurs in surface waters. Ocean acidification for surface waters can better be calculated using the assumption of equilibrium gas exchange and constant alkalinity, i.e., no equilibration with calcite. In this case (last column), pH declines more steeply without this buffering effect of carbonate minerals, i.e., the surface ocean is more sensitive to atmospheric composition changes on timescales shorter than ocean turnover (needed to equilibrate with sedimentary carbonates).

Case 4 The pH of a soda lake

Evaporative lakes are usually alkaline and contain soda minerals such as natron ($Na_2CO_3 \cdot 10H_2O$). The equilibrium reaction

$$Na_2CO_3 \cdot 10H_2O(s) \leftrightarrow 2Na^+ + CO_3^{2-} + 10H_2O$$

constrains the carbonate ion and sodium concentration of the lake. Calculate the pH of the alkaline lake knowing that the dissolved sodium concentration ($[Na^+]$) is 2.3 g kg^{-1}, that the lake water has a temperature of 25 °C and assuming equilibrium with the atmosphere, and a water density of 1 kg L^{-1}.

Table 6.2 Freshwater and ocean pH in equilibrium with calcite

	(P_{CO_2} µatm)	Freshwater pH with calcite (Eq. 6.23)	Ocean pH with calcite (Eq. 6.23)	Ocean pH with TA = 2750 µM
Glacial	180	8.43	8.34	8.37
Pre-industrial	270	8.31	8.22	8.24
Modern	420	8.18	8.09	8.09
2°C-2100	475	8.14	8.05	8.05
4°C-2100	800	8.00	7.90	7.86
Eocene	1000	7.93	7.84	7.77

Last column calculations with fixed TA rather than in equilibrium with calcite have been performed numerically

To solve this problem, we first identify the six unknowns species concentrations: $[H^+]$, $[OH^-]$, $[H_2CO_3]$, $[HCO_3^-]$, $[CO_3^{2-}]$ and $[Na^+]$. Consequently, we need six equations to solve the problem. The first four are the self-ionization of water (Eq. 6.6), the first and second acid-base equilibria (Eqs. 6.13 and 6.14) and Henry's Law combined with the P_{CO_2} (Eq. 6.10). The dissolved sodium concentration $[Na^+]$ is 0.1 mol L^{-1} and provides the fifth equation (mass balance). To close the system, we derive the charge balance:

$$[H^+] + [Na^+] = [OH^-] + [HCO_3^-] + 2[CO_3^{2-}] \tag{6.35}$$

For an alkaline lake, $[Na^+] \gg [H^+]$, and we can thus modify the charge balance and link it to the mass balance:

$$[Na^+] = [OH^-] + [HCO_3^-] + 2[CO_3^{2-}] = 0.1 \tag{6.36}$$

Since it is an open system in equilibrium with the atmosphere, we follow the approach for the previous case and use Eqs. 6.6, 6.26, and 6.27 to rewrite the charge balance (Eq. 6.36):

$$\frac{K_w}{[H^+]} + \frac{K_1 K_H \cdot P_{CO_2}}{[H^+]} + 2\frac{K_1 K_2 K_H \cdot P_{CO_2}}{[H^+]^2} = 0.1 \tag{6.37}$$

and obtain an equation with all terms known except for $[H^+]$.

Substituting the appropriate values for the constants,

$$\frac{10^{-14}}{[H^+]} + \frac{10^{-11.19}}{[H^+]} + 2\frac{10^{-21.53}}{[H^+]^2} = 0.1$$

and multiplying with $[H^+]^2$ we arrive at

$$10^{-14} \cdot [H^+] + 10^{-11.19} \cdot [H^+] + 10^{-21.23} = 10^{-1}[H^+]^2$$

Which after re-arrangement turns out to be a quadratic equation

$$[H^+]^2 - 10^{-10.19} \cdot [H^+] - 10^{-20.23} = 0$$

and results in a $[H^+]$ of 1.1×10^{-10} and thus a pH of 9.9. This estimate is approximate because of our neglect of non-ideal solution effects, i.e., using concentrations rather than activities.

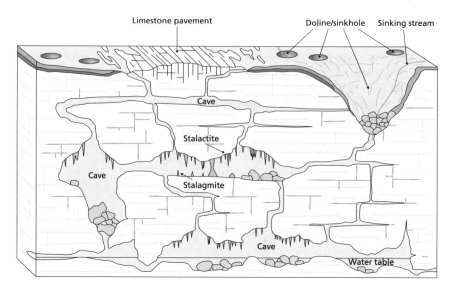

Fig. 6.2 Conceptual figure of karst system with dissolution at surface and precipitation in caves

Box 4 The Physical Chemistry of Karst

Rainwater entering soils becomes enriched in carbon dioxide from respiration of the primary producers (trees and plants), and the microbial degradation of soil organic matter. These higher P_{CO_2} levels cause dissolution of carbonate bed rocks, and consequently dolines and sinkholes are formed, and streams and rivers may go underground. Soil waters will eventually recharge subsurface water reservoirs. When the groundwater drains into stream or rivers exposed to air, carbon dioxide will escape from the water to the air to re-establish equilibrium. Consequently, the water becomes supersaturated with respect to calcium carbonates and travertine is formed. Similarly, when groundwater enters caves with lower P_{CO_2} levels, carbon dioxide will be transferred to the gas phase and stalagmites and stalactites may slowly precipitate (Fig. 6.2).

Consider a typical soil P_{CO_2} level of 2000 ppm (because of extensive root and microbial respiration). Using equation (6.34) we can estimate that soil water in equilibrium with calcite (carbonate bedrock) would have a pH of 7.73. To calculate the dissolved calcium concentration in the soil, we must re-arrange Eq. (6.33) to isolate $[Ca^{2+}]$ on the left-hand side.

$$[Ca^{2+}]^3 = P_{CO_2}\left[\frac{K_1 K_{sp} K_H}{4K_2}\right] \qquad (6.38)$$

Soil water recharging the groundwater would have a $[Ca^{2+}]$ of 0.81 mM (32.5 mg L^{-1}) and TA and DIC values of 1.63 mM each. If we assume that

$[Ca^{2+}]$, TA and DIC are transported conservatively, i.e., they do not change during transport, and that downstream cave/stream waters re-equilibrate with air with a P_{CO_2} level of 420 ppm, we can calculate the full carbon dioxide system in equilibrium. Cave and stream re-equilibrate with air and loose about 0.66 mM DIC. The equilibrium pH, TA and $[Ca^{2+}]$ would be 8.18, 0.98 mM and 0.48 mM (19.3 mg L^{-1}), respectively, because of calcium carbonate precipitation induced by the carbon dioxide transfer from water to air.

	Soil water (P_{CO_2} = 2000 ppm)	Cave/stream water (P_{CO_2} = 420 ppm)
pH	7.73	8.18
$[Ca^{2+}]$ (mM)	0.81	0.48
DIC (mM)	1.63	0.97
TA (mM)	1.63	0.98
$[HCO_3^-]$ (mM)	1.62	0.97
$[CO_3^{2-}]$ (μM)	4.1	6.8

Chapter 7
Redox Equilibria

Abstract This chapter focuses on equilibria involving electron transfers (redox reactions). A systematic approach to identify electron transfers and to balance redox reactions is introduced. Redox potential and the Nernst equation are linked to the Gibbs free energy.

Keywords Electron acceptor · Electron donor · Redox potential · Nernst equation · Oxidation state

7.1 Introduction

Redox reactions involve the transfer of electrons from an electron donor (reductant) to an electron acceptor (oxidant). The electron donor/reductant is then oxidized, while the electron acceptor (oxidant) is reduced. This 'reductant is oxidized' and 'oxidant is reduced' terminology could be seen as confusing, and for clarity we will avoid the use of the terms reductant and oxidant and limit the use of the terms oxidation and reduction. Electron transfers are linked to energy transfers and provide the basis for most life on Earth. Heterotrophs, including humans, utilize organic matter (an electron donor) and combine it with an electron acceptor (oxygen) to respire and release the energy for maintenance and growth. Some microbes (e.g. chemolitho(auto)trophs) use minerals such as pyrite (FeS_2) or reduced gases such as methane (CH_4) as an electron donor to generate energy upon electron transfer to an acceptor (oxygen, metal oxides or sulfate). The exposure of reduced substances (organic matter, iron sulfide) to atmospheric oxygen (oxidative weathering) causes not only transfer of electrons (oxidation), but also release of protons. An extreme example is acid mine drainage: the oxidation of metal sulfides causes proton release with acidic conditions as a result. Pyrite (FeS_2) is a major constituent of mine waste tailings and reacts with oxygen:

$$4FeS_2 + 15O_2 + 10H_2O \rightarrow 4FeOOH + 8SO_4^{2-} + 16H^+$$

J. J. Middelburg, *Thermodynamics and Equilibria in Earth System Sciences: An Introduction*, SpringerBriefs in Earth System Sciences,
https://doi.org/10.1007/978-3-031-53407-2_7

This reaction generates 4 protons per mol pyrite oxidized and leads to acidification.

7.2   Identifying Electron Transfer and Balancing Redox Reactions

Redox reactions change the oxidation state of substances. An *oxidation state, or oxidation number,* is a hypothetical charge of an atom, if all its bonds to other atoms were fully ionic. It quantifies the excess or shortage of electrons. Oxidation numbers are integers because only discrete electrons can be transferred. The rules for assigning oxidation numbers are as follows:

1. An atom in its elemental state has an oxidation number of 0: e.g., Na, O_2, Ar
2. An atom in a monoatomic ion has an oxidation number equal to the charge of the ion: e.g., Ca^{2+}, Fe^{3+}, F^-
3. Oxygen has an oxidation state of -2, except when bonded to an oxygen or fluor atom.
4. Hydrogen has an oxidation state of $+1$, except when bonded as a hydride (-1) to metals.
5. Alkaline and earth alkaline metals (columns 1 and 2 of periodic table) have an oxidation state of $+1$ and $+2$, respectively, and the halogens (F, Br, Cl, I) have an oxidation state of -1, except when bonded to oxygen or nitrogen.
6. The sum of oxidation numbers is equal to the net charge for a polyatomic ion.

For instance, in sulfuric acid, H_2SO_4, sulfur has an oxidation number of $+6$, while nitrogen oxidation numbers in nitrate (NO_3^-) and ammonium (NH_4^+) are $+5$ and -3, respectively. Assigning oxidation states to sulfur is challenging as it may be -2, -1, 0, $+3$, $+4$ and $+6$, depending on the molecular setting. It is important to recall that these oxidation numbers are hypothetical because one would infer that the charge of the two sulfur atoms in thiosulfate $(S_2O_3^{2-})$ is $+2$, while in fact one sulfur has a charge of -1 and the other $+5$. Some common (Fe, Mn, Ti-bearing) minerals have elements with mixed oxidation number: e.g., in magnetite, Fe_3O_4, two of the iron atoms are $+3$ and the other is $+2$.

Balancing reactions involving electron transfers requires a systematic approach involving two steps, first the identification of the electron donor and acceptor, hence the electrons transferred, and then the balancing of the elements. There are two methods: the half-reaction, or ion–electron method, in which the oxidation and reduction processes are separated and each balanced before re-combining them again, and the oxidation state or whole-reaction method. The latter involves the following steps:

(1)   Assign oxidation state numbers to each of the atoms in the equation and write the numbers above the atom.
(2)   Identify the atoms that donate or accept electrons, i.e., that increase or decrease in oxidation number.

(3) Identify the electron donating and electron accepting compounds on the left-hand side and quantify the number of electrons they donate/accept.
(4) Balance the electron donating and accepting compounds on the left-hand side
(5) Balance all elements, except O and H, on both sides of the equation
(6) Balance oxygen on both sides of the equation through addition of H_2O
(7) Balance hydrogen on both sides of the equation through addition of H^+

To illustrate this method, consider the following reaction that may occur in surface sediments or in groundwaters:

$$FeS + NO_3^- \rightarrow FeOOH + SO_4^{2-} + N_2$$

First, we identify the oxidation numbers (oxygen and hydrogen are -2 and $+1$, respectively)

$$\begin{array}{ccccc} +2 -2 & +5 & +3 & +6 & 0 \\ FeS + & NO_3^- \rightarrow & FeOOH + & SO_4^{2-} + & N_2 \end{array}$$

and recognize that Fe donates one electron to go from $+2$ to $+3$, S donates 8 electrons to go from -2 to $+6$ and N accepts 5 electrons to go from $+5$ to 0. Accordingly, FeS donates 9 electrons (1 from Fe and 8 from S) and NO_3^- accepts 5 electrons (N from $+5$ to 0). Balancing the electron donor and acceptor on the left-hand side thus implies that 5 FeS react with 9 NO_3^-.

$$5FeS + 9NO_3^- \rightarrow FeOOH + SO_4^{2-} + N_2$$

The rest of the balancing is simply a book-keeping, first, we balance Fe, S and N:

$$5FeS + 9NO_3^- \rightarrow 5FeOOH + 5SO_4^{2-} + 4.5N_2$$

Next, we balance oxygen atoms by adding three molecules of H_2O, in this case to the left-hand side:

$$5FeS + 9NO_3^- + 3H_2O \rightarrow 5FeOOH + 5SO_4^{2-} + 4.5N_2$$

And finally, we balance hydrogen atoms by adding one proton, in this case to the right-hand side:

$$5FeS + 9NO_3^- + 3H_2O \rightarrow 5FeOOH + 5SO_4^{2-} + 4.5N_2 + H^+$$

7.3   Redox Potentials and the Nernst Equation

Electron transfers are associated with energy transfer. Electrochemical cells (e.g., a battery) convert Gibbs free energy of a reaction into electrical energy and vice versa. Galvanic cells make use of spontaneous chemical reactions to generate electricity, while electrolysis requires electrical energy to drive a non-spontaneous reaction.

Consider a strip of zinc metal in an aqueous solution of copper sulfate, the strip will become darker because of copper precipitates and the blue color solution will fade because of the spontaneous reaction consuming the colored Cu^{2+}

$$Zn(s) + Cu^{2+}(aq) \rightarrow Zn^{2+}(aq) + Cu(s) \tag{7.1}$$

Zn donates two electrons which are accepted by Cu^{2+} and the energy released goes into heat. If the electron donating (oxidation) and accepting (reduction) part of the reaction are physically separated but connected via a wire and salt bridge (an ionic solution allowing migration of cations and anions) to maintain charge balance, an electrical current will flow through the wire because of a difference in potential (E, expressed in V(olt)). Electrochemists split redox reactions in two half-reactions, or half-cells, one for the oxidation and one for the reduction process. For reaction 7.1, these half-reactions are:

$$Zn(s) \rightarrow Zn^{2+}(aq) + 2e^- \text{ for the oxidation (electrons donating reaction)} \tag{7.2a}$$

and

$$Cu^{2+}(aq) + 2e^- \rightarrow Cu(s) \text{ for the reduction (electrons accepting reaction).} \tag{7.2b}$$

The potentials for these half reactions have been measured relative to that of protons/hydrogen gas conversions which has been set at 0 V for both reduction and oxidation:

$$2H^+(aq) + 2e^- \rightarrow H_2(g), \text{ reduction, electrons accepting reaction} \tag{7.3a}$$

$$H_2(g) \rightarrow 2H^+(aq) + 2e^-, \text{ oxidation, electrons donating reaction} \tag{7.3b}$$

The *standard reduction potentials* E^0 (with electrons on the left-hand side and at SATP) have been tabulated and can be used to calculate the redox potential of reactions (Table 7.1). Oxidation potentials are the reverse of reduction potentials. For instance, the standard reduction potential for Zn and Cu are:

$$Zn^{2+} + 2e^- \rightarrow Zn \quad E^0 = -0.76 \text{ V} \tag{7.4a}$$

$$Cu^{2+} + 2e^- \rightarrow Cu \quad E^0 = +0.34\,V \tag{7.4b}$$

The overall potential for the complete reaction 7.1 is then $-(-0.76) + 0.34 =$ 1.1 V, resulting in a positive voltage as it should be for a spontaneous reaction. Redox potentials are intensive properties, i.e., independent of the amount, and additive.

The standard redox potential E^0 is related to the standard Gibbs free energy via the *Nernst expression*:

$$\Delta_r G^o = -nFE^0 \tag{7.5}$$

where $\Delta_r G^o$ is the standard Gibbs free energy of a reaction, n is the number of electrons transferred and F is the Faraday constant (the electrical charge of 1 mol of electrons: 96,500 C (C) per mol e^-). Recall from physics class that $1\,J = 1\,V\cdot C$. The minus sign in Eq. (7.5) is needed because spontaneous reactions have a negative Gibbs free energy and a positive redox potential. Accordingly, we can calculate the Gibbs free energy of reaction 7.1:

$$\Delta_r G^o = -nFE^0 = -2 * 96500 * 1.1 = -212kJ$$

Table 7.1 Standard reduction potentials at 25 °C

Reduction Half-Reaction		E^o (V)
Co^{3+} (aq) + e$^-$	→ Co^{2+} (aq)	1.92
H_2O_2 (aq) + 2 H$^+$ (aq) + 2 e$^-$	→ 2 H$_2$O (l)	1.77
Ce^{4+} (aq) + e$^-$	→ Ce^{3+} (aq)	1.72
Mn^{3+} (aq) + e$^-$	→ Mn^{2+} (aq)	1.5
$Cr_2O_7^{2-}$ (aq) + 14 H$^+$ (aq) + 6 e$^-$	→ 2 Cr^{3+} (aq) + 7 H$_2$O (l)	1.36
O_2 (g) + 4 H$^+$ (aq) + 4 e$^-$	→ 2 H$_2$O (l)	1.23
Ag^+ (aq) + e$^-$	→ Ag (s)	0.80
Fe^{3+} (aq) + e$^-$	→ Fe^{2+} (aq)	0.77
Cu^{2+} (aq) + 2 e$^-$	→ Cu (s)	0.34
Cu^{2+} (aq) + e$^-$	→ Cu$^+$ (aq)	0.16
$2H^+$ (aq) + 2 e$^-$	→ H$_2$ (g)	0.0
Pb^{2+} (aq) + 2 e$^-$	→ Pb (s)	−0.13
Sn^{2+} (aq) + 2 e$^-$	→ Sn (s)	−0.14
Co^{2+} (aq) + 2 e$^-$	→ Co (s)	−0.28
Cd^{2+} (aq) + 2 e$^-$	→ Cd (s)	−0.40
Fe^{2+} (aq) + 2 e$^-$	→ Fe (s)	−0.44
Zn^{2+} (aq) + 2 e$^-$	→ Zn (s)	−0.76
Mn^{2+} (aq) + 2 e$^-$	→ Mn (s)	−1.17
Mg^{2+} (aq) + 2 e$^-$	→ Mg(s)	−2.36

The standard redox potential is also related to the equilibrium constant because $\Delta_r G^o$ is linked to K: (Eq. 5.10):

$$K = \exp\left(\frac{-\Delta_r G^o}{RT}\right) = \exp\left(\frac{nFE^0}{RT}\right) \tag{7.6}$$

.

Redox potentials, like Gibbs free energy changes, depend on temperature and the composition of the reaction mixture. For Gibbs free energy, we have (Eq. 5.9):

$$\Delta_r G = \Delta_r G^o + RT \ln Q$$

where Q is the reaction quotient. Since $\Delta_r G^o = -nFE^0$ and $\Delta_r G = -nFE$, we arrive at

$$-nFE = -nFE^0 + RT \ln Q$$

Dividing by $-nF$, this results in the *Nernst equation*

$$E = E^0 - \frac{RT}{nF} \ln Q = E^0 - \frac{2.303RT}{nF} \log Q \tag{7.7}$$

Recall from math class that $\ln 10 = 2.303 \log 10 = 2.303$.
At 25 °C, the term $\frac{2.303RT}{F}$ has a value of 0.0592 V and thus

$$E = E^0 - \frac{0.0592}{n} \log Q. \tag{7.8}$$

Returning to our initial reaction (7.1), we consider the case that $\left[Zn^{2+}\right]$ has a concentration of 0.5 M and $\left[Cu^{2+}\right] = 0.01$ M. The potential would then be

$$E = E^0 - \frac{0.0592}{n} \log Q = 1.1 - \frac{0.0592}{2} \log \frac{\left[Zn^{2+}\right]}{\left[Cu^{2+}\right]} = 1.05 \text{ V}.$$

The redox reaction is favorable.

We conclude with articulating the internal consistency and interchangeability of equilibrium constants between different branches of chemistry. Equilibrium constants can be derived from:

thermochemistry, i.e., ΔG^o_f data obtained on pure phases (Eq. 5.9): $K = \exp\left(\frac{-\Delta G^o_r}{RT}\right)$

the composition of solutions at equilibrium (Eq. 5.4) $K = \frac{C^\gamma_{eq} D^\delta_{eq}}{A^\alpha_{eq} B^\beta_{eq}}$

and redox potentials (Eq. 6): $K = \exp\left(\frac{nFE^0}{RT}\right)$.

In the absence of uncertainties, the equilibrium constants should be identical.

Literature and Further Reading

Atkins, P, de Paula, J., Keeler J., 2006. Atkin's Physical Chemistry, 8nd Edition, W.H. Freeman and Company, 1085 p.
   Well-written, accessible physical chemistry book with 240 pages of thermodynamics.
Ball D.W., 2003. Physical Chemistry, 2nd edition, Brooks/Cole, Thomson, 857 p.
   Calculus based physical chemistry textbook including basic treatment of thermodynamics.
Butler J.N., 1991. Carbon Dioxide Equilibria and Their Applications. Routledge, New York, 272 p. https://doi.org/10.1201/9781315138770
   Accessible book on carbon dioxide equilibria.
Feynman, R. P., Leighton R.B., Sands, M. 1963. The Feynman Lectures on Physics. https://www.feynmanlectures.caltech.edu/I_toc.html
   Chapter 44 gives an instructive,concise treatment of the laws of thermodynamics.
Garrels, R.M., Christ, C.L., 1965. Solutions, Minerals and Equilibria. W. H. Freeman, San Francisco, 450 p
   This seminal book links thermodynamics, equilibria, minerals, and solutions.
Kleidon, A., 2016. Thermodynamic Foundations of the Earth System. Cambridge University Press, 379 p. https://doi.org/10.1017/CBO9781139342742
   This book provides a thorough treatment of how thermodynamics applies to Earth system processes.
Lowrie, W., 2011. A Student's Guide to Geophysical Equations. Cambridge University Press, 281 p. https://doi.org/10.1017/CBO9780511794438
   Provides nice geophysical examples and inspiration for Box 2.
McMurry, J.E., Fay R.C., 2012. Chemistry, 6th edition, 1075 p.
   Introduction to chemistry, includes thermodynamics, gases and equilibria.
Middelburg, J.J., 2019. Marine Carbon Biogeochemistry, A primer for Earth System Scientists, Springer, https://doi.org/10.1007/978-3-030-10822-9

J. J. Middelburg, *Thermodynamics and Equilibria in Earth System Sciences: An Introduction*, SpringerBriefs in Earth System Sciences, https://doi.org/10.1007/978-3-031-53407-2

Chapter 5 provides a more detailed treatment of carbon dioxide equilibria in seawater.

Price, G., 2019. Thermodynamics of Chemical Processes. Oxford University Press, 128 p.

An undergraduate level concise treatment of thermodynamics of chemical processes.

Stumm, W., Morgan, J.J., 1996, Aquatic Chemistry, Chemical Equilibria and Rates in Natural Waters. 3rd Edition, John Wiley & Sons, Inc., New York.

Comprehensive treatment of aquatic chemistry, including homogenous and heterogenous equilibria.

Index

© The Editor(s) (if applicable) and The Author(s) 2024
J. J. Middelburg, *Thermodynamics and Equilibria in Earth System Sciences: An
Introduction*, SpringerBriefs in Earth System Sciences,
https://doi.org/10.1007/978-3-031-53407-2

Printed in the United States
by Baker & Taylor Publisher Services